Dimitri Plemenos and Georgios Miaoulis (Eds.)

Intelligent Computer Graphics 2011

Studies in Computational Intelligence, Volume 374

Editor-in-Chief

Prof. Janusz Kacprzyk
Systems Research Institute
Polish Academy of Sciences
ul. Newelska 6
01-447 Warsaw
Poland
E-mail: kacprzyk@ibspan.waw.pl

Dimitri Plemenos and Georgios Miaoulis (Eds.)

Intelligent Computer Graphics 2011

 Springer

Editors

Dimitri Plemenos
University of Limoges
44, rue Antoine Bourdelle
87000 Limoges
France
E-mail: plemenos@numericable.com

Georgios Miaoulis
Technological Education Institute of Athens
Department of Computer Science
Ag. Spyridonos
Egaleo, 122 10 ATHENS
Greece
E-mail: gmiaoul@teiath.gr

ISBN 978-3-642-22906-0   e-ISBN 978-3-642-22907-7

DOI 10.1007/978-3-642-22907-7

Studies in Computational Intelligence   ISSN 1860-949X

Library of Congress Control Number: 2011936703

© 2012 Springer-Verlag Berlin Heidelberg

Typeset & Cover Design: Scientific Publishing Services Pvt. Ltd., Chennai, India.

Printed on acid-free paper

9 8 7 6 5 4 3 2 1

springer.com

Preface

Intelligent techniques are used since several years in various research areas using computer programs. The purpose of using intelligent techniques is generally to optimise the processing time, to find more accurate solutions, for a lot of problems, than with traditional methods, or simply to find solutions in problems where traditional methods fail. By "intelligent techniques" we mean above all Artificial Intelligence based techniques but not only. In some cases, simple human intelligence may help invent problem adapted new processing methods which greatly improve existing ones. In Computer Graphics, the use of intelligent techniques started more recently than in other research areas. However, during these last two decades, the use of intelligent Computer Graphics techniques is growing up year after year and more and more interesting techniques are presented in this area.

The purpose of this volume is to present current work of the Intelligent Computer Graphics community, a community growing up year after year. This volume is a kind of continuation of the previously published Springer volumes "Artificial Intelligence Techniques for Computer Graphics" (2008), "Intelligent Computer Graphics 2009" (2009) and "Intelligent Computer Graphics 2010" (2010).

What is Intelligent Computer Graphics? It is a set of Computer Graphics problems whose solution is strongly improved by the use of intelligent techniques. These techniques are mainly based on Artificial Intelligence. So, in Declarative scene Modelling, problem resolution, constraint satisfaction and machine-learning techniques are used. In scene understanding, as well as in improved Monte-Carlo Radiosity, heuristic search techniques allow to improve solutions. In virtual world exploration, efficient camera movement is achieved by strategy games techniques. In behavioural animation, multi-agent techniques, as well as evolutionary algorithms are currently used.

However, it is obvious that techniques based on Artificial Intelligence cannot resolve all kinds of problems. In some cases, the use of specific Artificial Intelligence techniques may become too heavy and even inefficient, while, sometimes, simple human intelligence, easy to implement, can help find interesting solutions in cases where traditional Computer Graphics techniques, even combined with Artificial Intelligence ones, cannot propose any satisfactory solution. Such a case is the one of visual scene understanding, where it is sometimes easy to know what kind of view is expected by the user. Another case where the use of simple human intelligence is often requested is data visualisation, when a little bit of imagination can give interesting results.

During a long time, Artificial Intelligence techniques remained unknown and unused for Computer Graphics researchers, while they were already used in other graphic processing areas like image processing and pattern recognition. We think that the 3IA International Conference on Computer Graphics and Artificial Intelligence, organised since 1994, grandly contributed to convince many Computer Graphics researchers that intelligent techniques may allow substantial improvements in a lot of Computer Graphics areas. Nowadays, more and more researchers in Computer Graphics all over the world are interested in intelligent techniques. We think that the main contribution of techniques issued from Artificial Intelligence is to allow invention of completely new methods in Computer Graphics, often based on automation of a lot of tasks assumed in the past by the user in an imprecise, often inefficient and (human) time consuming manner.

For Computer Graphics researchers it is important to know how the use of intelligent techniques evolves every year and how new intelligent techniques are used in new areas of Computer Graphics year after year.

When the 3IA International Conference on Computer Graphics and Artificial Intelligence was first created by Dimitri PLEMENOS in 1994, its purpose was to put together Computer Graphics researchers wishing to use Artificial Intelligence techniques in their research areas, in order to create emulation among them. Nowadays, seventeen years after the first 3IA International Conference (3IA'94), the number of Computer Graphics researchers using Artificial Intelligence techniques became very important. Thus, an additional purpose of the 3IA Conference is to keep researchers informed on the existence of new intelligent methods, and even of corresponding software, for specific areas of Computer Graphics.

This volume contains selected extended papers from the last 3IA Conference (3IA'2011), which has been held in Athens (Greece) in May 2011. This year papers are particularly exciting and concern areas like virtual reality, artificial life, data visualization, games, global illumination, point cloud modelling, declarative modelling, scene reconstruction and many other very important themes.

We think that in Computer Graphics still exist a lot of areas where it is possible to apply intelligent techniques. So, we hope that this volume will be interesting for the reader and that it will convince him (her) to use, or to invent, intelligent techniques in Computer Graphics and, maybe, to join the Intelligent Computer Graphics community.

Dimitri Plemenos
Georgios Miaoulis

Contents

Detecting Visual Convergence for Stochastic Global Illumination

Christophe Renaud[1], Samuel Delepoulle[2], and Nawel Takouachet

[1] LISIC, BP 719 - 62228 Calais cedex France
   renaud@lisic.univ-littoral.fr
[2] LISIC, BP 719 - 62228 Calais cedex France
   delepoulle@lisic.univ-littoral.fr

Abstract. Photorealistic rendering, based on unbiased stochastic global illumination, is now within reach of any computer artist by using commercially or freely available softwares. One of the drawbacks of these softwares is that they do not provide any tool for detecting when convergence is reached, relying entirely on the user for deciding when stopping the computations. In this paper we detail two methods that aim at finding perceptual convergence thresholds for solving this problem. The first one uses the VDP image quality measurement for providing a global threshold. The second one uses SVM classifiers which are trained and used on small subparts of images and allow to take into account the heterogeneity of convergence through the image area. These two approaches are validated by using experimentations with human subjects.

1 Introduction

High quality synthetic images are available for several years thanks to the use of photorealistics lighting methods. Those are mainly based on statistical sampling of the lighting paths that join up the observer's eye or the camera to the virtual objects and light sources [Kaj86]. Increasing the number of lighting paths allows the methods to smoothly converge to the correct image and/or to the desired numerical lighting values. Because the convergence rate of such processes is generally low, most of the researches in photorealistic rendering have been focused on the ways of accelerating this convergence. Powerfull algorithms have been designed such as the Metropolis light transport [VG97] and are now commercially[Nex] or freely [Lux] available to computer graphics designers.

The main drawback of thoses methods as highlighted by the final users is the difficulty to determine when stopping the computation process. Indeed the stochastical nature of the algorithms ensures that they converge toward the true solution. But it produces numerical noise that is visually perceptible as color high frequencies. This noise disappears progressively as the computation process converges but it is difficult to determine when convergence is achieved. Numerical accuracy can be checked for this purpose but its results are generally much more accurate than those that could be visually used. Indeed the human visual system, that is the final

D. Plemenos and G. Miaoulis (Eds.): Intelligent Comp. Graphics 2011, SCI 374, pp. 1–17.
springerlink.com

target for the images, is far as accurate as the numerical accuracy of any computer. As a consequence the computation times are generally higher than those that would be required if more user-dependent visual thresholds were used.

At our knowledge none of the available photorealistic softwares provides any efficient stopping strategy. As a consequence computer graphics artists use these softwares in a compute-and-test approach: they run their software for a given computation time, save the results and look at the image. When this image is considered to be noisy they resume the computation for a new time period. And so on until their image can be considered as converged. This way of doing is clearly far to be pleasant nor efficient: convergence rate is generally different between each part of any image according to the local properties of the virtual scene (lighting, materials of the objects, ...). Resuming the computation for the entire image each time even for those parts that appear as visually converged increases considerably the computation times. Conversely managing manually the computation for each part of the image is fussy and difficult.

In this paper we are interested in studying some ways of automating the research of the convergence threshold of such lighting simulations. Because the details of the software code are generally not available, we investigate the ways of using only the results provided as outputs by these softwares, mainly RVB images. Furthermore we focus on the research of visual thresholds that should be more efficient for computer graphics applications by exploiting the properties and the limits of the Human Visual System (HVS).

In the next section we describe some of the works that have been done in studying the HVS and their applications in realistic rendering. Section 3 is devoted to the presentation of experimentations that aim at acquiring real data about the perception of noise by human subjects. We present then the use of this data through two approaches whose goal is to provide efficient ways of stopping lighting calculations when perceptual convergence is reached. The first one in section 4 uses the Visual Difference Predictor image quality measurement. The second one (section 5) uses Support Vector Machines which are part of classification methods used in A.I. Some perspectives are given in section 6.

2   Section Heading

Considerable research efforts have been devoted to understanding and simulating the Human Visual System behavior (HVS). These researches highlighted the properties of the HSV and showed its limits.

From these studies two main kinds of perceptual models have been designed. At one hand, the models that define perceptual quality metrics which are then used for measuring simularities and differences between images. The Visible Difference Predictor (VDP) [Dal93] and the Sarnoff's Visual Discrimination Model (VDM) [Sar97] are well-known models of this kind. At the other hand some perceptual models are based on visual attention which is the process of selecting a portion of the available visual information for localization, identification or understanding of objects in an environment. Several models for driving the visual attention have been proposed [TG80] [KU85][NIK01][Itt00] which output is a map of saliency which simulates the eye movements between regions or objects of interest in images.

These models have been used by computer graphics researchers in order to define and study some new perceptually-based techniques for realistic rendering. Yee [Yee04] has proposed an abridged version of the VDP in the same way as Ramasubramanian [RPG99] in which they drop the orientation computation when calculating spatial frequencies and extend the VDP by including the color domain in computing the differences. The resulting model is fastest than the original VDP which is important when using it on a video sequence. Farrugia and Peroche [FP04] proposed a perceptually-based rendering method in which the rendering accuracy needed per pixel is adjusted according to a perceptual adaptive metric based on the Multi-scale Model of Adaptation and Spatial Vision [PFFG98]. This allows to improve rendering time by saving calculations in some regions without the viewer being able to detect any differences between the refined image and the one computed with a standard global illumination method. Finally both [Mys98] and [Tak09] use the VDP in order to provide a quantitative measure of perceptual convergence.

3   Experimental Data Acquisition

Because we are interested in using some perceptual stopping condition we need to get some information about how and where noise is visible in any image. At our knowledge not any model nor data are available for this purpose. We thus rely on real data that must be acquired through experiments with real people and images sets. Once available these information can be used when validating the stopping methods we will describe in sections 4 and 5.

The main problem in noise detection is to clearly explain to experimentation subjects what noise is and what they should perceive in the different images. Then even with a clear understanding of what should be noted in any image, it is often not easy to decide whether any part of an image is noisy if no information about how the converged image should appear. For these reasons it is necessary to give very simple instructions to the subjects and to provide them with a reference image that is supposed to be free of noise. This reference image can be computed thanks to a purely numerical threshold.

3.1   Acquiring a Per Image Threshold

Some methods [TDR07][Mys98] propose to stop the stochastic lighting algorithm by analyzing noise in the entire image. Validating their results thus requires to get experimental noise detection thresholds for several different all-round images. For each of those images, lighting computation is performed according to sampling steps: the chosen global illumination algorithm is run for a specified number of light path samples per pixel or for a given computation time. The corresponding image is recorded and the algorithm is resumed for a new step and so on until the lighting computation converges to a noise free image. This approach is run for several different images I_n and provides us with a set of images $I_n^{(p)}$, n being the image index and p the value of the sampling level of image I_n. According to the way the sampling levels are defined, p could be a number of light path samples per pixel or the time required for computing the corresponding image. The last image of each set, $I_n^{(max)}$, is called the *reference image* and is supposed to be free of visual noise.

3.1.1 General Approach

Those images are then used into a dedicated application which purpose is to estimate a threshold for noise perception for any of the I_n images. More specifically we want to estimate the average value of p for which most of the users do not perceive any noise in the corresponding image. A way to perform this task is to provide the users a pair of images: one is a reference image $I_n^{(max)}$ and the other is one of the $I_n^{(p)}$ intermediate images. Then the user is asked whether these images are identical or not. Performing this task for all the images I_n and for several users should allow us to get a good estimate of the visual noise perception threshold for each image. These thresholds can then be used for evaluating the results of the methods of noise detection that will be described in the next sections.

3.1.2 The Experiment

From a practical point of view this approach is far to be user-friendly: the number of images samples $I_n^{(p)}$ is generally high for each image and for generality purpose we need to use several I_n test images. Thus any user has to estimate the quality of a very large number of couples of images which requires long experimentation times and decreases the accuracy of the answers.

We thus divide up the approach in two steps: the first one aims at finding an interval of sampling levels for each image, in which 50% of the users did not perceive any noise; then some sampling levels are selected into this interval and are used for improving the search of the threshold.

In first fast step we presented successively couples of images $(I_n^{(max)}, I_n^{(p)})$, $p \in [50..10000]$ light path samples/pixel with a step of 50 samples/pixel. Subjects were asked to stop the display of couples when they did not perceive any difference between the two images. In order to reinforce statistical results, we shown the series of images in ascending and descending order. We then deduce from this experiments the interval of sampling levels in which 50% of the subjects do not perceive difference between the reference image and the noisy ones. This interval is $[100, 2500]$ for the different images that were used (see figure 1).

In the second step we selected 7 regularly spaced images in the interval. These images are used with the reference one for displaying couple of images and asking the subjects whether these images are identical or not (see figure 2). Five displays of the same couple is done for each different scenes. The images were shown simultaneously as long as necessary for the subject to take the decision to answer. The subject gives his answer by pressing a button on the computer interface with the mouse. The order of presentation is counterbalanced with a pseudo-randomized schedule, furthermore the reference image is shown randomly at the left or at the right of the target image. This procedure is conform to the measure of the differential threshold with the method of constant stimuli.

Fig. 1 The eight converged images used during the experiments: four scenes were used with two different light intensities for each

For the two steps, the subjects of the experiment were 16 undergraduate students (14 males and 2 females). The average age is 21.0 with a standard deviation of 3.35. Subjects were placed at a distance of 0.5 m of the display (19 flat panel display at resolution $1280 \times 1024, 300 \, cd/m^2$).

After the second step we get more accurate measures about how much subjects perceive some noise at each of the 7 chosen sampling levels. In order to now get the value of the thresholds we assume that the probability to perceive a difference between any image $I_n^{(p)}$ and its reference one follows a sigmoid law; we thus can approximate the results with a logit function which general expression is:

$$p(x) = \frac{c.e^{a.x+b}}{1 + e^{a.x+b}} \qquad (1)$$

Table 1 shows the value of the coefficients of equation 1 computed by a logit regression, r being the regression coefficient. The values obtained for r are closed to 1 which highlights the good correlation between the experimental data. According to these functions we are now able to compute a perceptual threshold for an entire image with any value of confidence. These thresholds will be used in section 4 during the validation stage of the corresponding approach.

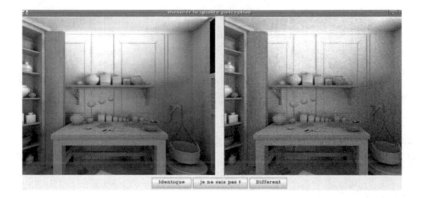

Fig. 2 A view of the interface used for the second step of the thresholds acquisition

Table 1 Estimates of the coefficients of a logit function for each of the 8 scenes

scene \ Coeff.	a	b	c	r
BAR 1	-0,00200	4,007	1,010	-0,940
BAR 2	-0,00178	4,558	1,010	-0,907
CUBE 1	-0,00226	2,677	1,010	-0,841
CUBE 2	-0,00186	3,489	1,010	-0,873
OCULIST 1	-0,00233	5,415	1,010	-0,940
OCULIST 2	-0,00178	5,474	1,010	-0,767
PLANTS 1	-0,00185	2,743	1,010	-0,797
PLANTS 2	-0,00173	2,457	1,010	-0,752

3.2 Acquiring Thresholds for Sub-images

By considering the entire image for noise thresholding we assume that noise is equally distributed through any image. The direct consequence of such assumption is that sampling has to be improved equally for each pixel of the images. However noise is generally not homogeneously distributed, due to different lighting, perceptual or geometrical parameters (shadows, textures, geometric complexity, ...). Noise thresholds are thus different for each location in each image and the use of a global threshold reduces the computation efficiency by requiring the same number of samples to be computed for each pixel of an image.

However acquiring accurate data about the noise thresholds for individual parts of an image is much more complex for both the acquisition application and the subjects. In order to be able to reach this goal we first propose to set some limits to the locations where researching the noise thresholds. For this purpose any experimentation image is cut into a regular grid of nonoverlapping subimages of size 128×128

pixels. For our 512×512 test images we thus get 16 different subimages. This size has been chosen for allowing the subjects to easily watch to the corresponding image area[1].

We thus defined a very simple protocol in which pairs of images are presented to the observer. Similarly to section 3.1, one of this image is called the *reference image* and has been computed with $N_r = 10.000$ samples per pixel. The second image so called the *test image* is built as a stack of images, from very noisy ones above to converged ones below: by calling N_i the number of samples in the stack's image i, with $i = 1$ at the top of the stack and $i = max = N_r$ at its bottom, we thus ensure the property $\forall i \in [O, max[, N_i < N_{i+1} \leq N_r$. Each of these images are opaque and virtually cut into nonoverlapping subimages of size 128×128. For our 512×512 test images we thus get 16 different subimages for each of the stack's images.

During the experiments the observer is asked to modify the quality of the noisy image by pointing the areas where differences are perceived between the current image and its reference one. Each point-and-clic operation then causes the selection and the display of the corresponding $i + 1$ level subimage thus visually reducing the noise in this image's subpart. This operation is done until the observer considers that the two images are visually identical. Note that for reducing experiment artefacts this operation is reversible meaning that an observer is able to go down or up into the noisy images stack. The pair of images that is presented to the observer is chosen randomly but we ensure that each pair will be presented two times. Obviously the subimage grid is not visible and all the observers experimented in the same conditions (same display with identical luminance tuning, same illumination conditions, ...).

4579	6631	5818	5906
3271	2222	2261	2009
2400	2324	2098	2735
2955	2344	2002	2190

Fig. 3 An example of the subimages grid: the 512×512 image is divided into sixteen 128×128 non overlapping subimages. The grid has been drawn on the left image and the numbers in the right grid represent the average number of samples required for 95% of the observers to consider that the corresponding subimage is not noisy

---

[1] A smaller subimage size can be used but should increase the complexity for users to adjust the noise threshold using our protocol.

The results have been recorded for 33 different observers and we computed the average number of samples $\tilde{N}$ that are required for each subimage to be perceived as identical to the reference one by 95% of the observers. We got experimentally $\tilde{N} \in [1441, 6631]$ with often large differences between subimages of the same image (see figure 3).

4   Using the VDP

4.1   Principle

The principle of our approach is to compute incrementally the converged image by computing at step p an image $I^{(p)}$ with $p \times N$ light samples and by adding N light samples to the next image $I^{(p+1)}$ whether image $I^{(p)}$ still highlights visual noise. The value of N is chosen by the user and we used 100 in our experimentations with a Path Tracing method.

Deciding whether noise is visible is performed through the use of the VDP. However this method computes a visible distance between two images. Its use in our approach thus requires a second image that could be used as a reference one. The best choice would be the final image $I^{(max)}$; obviously this one is not available.

Myszkowski [Mys98] used a similar approach and proposed a method for estimating the perceptual differences between the current rendering ($I^{(p)}$) and the fully converged image ($I^{(max)}$) by using the following relation:

$$VDP(I^{(p)}, I^{(max)}) \approx VDP(I^{(p)}, I^{(\frac{p}{2})})$$

However this requires to store every image $I^{(p)}$ since it could be used later in the process. Furthermore as will be shown later this approach seems to overestimate the value of the thresholds. Consequently convergence is decided later than it could be done with a better threshold.

In [Tak09] we proposed to use the first noisy image $I^{(1)}$ as the reference one. The idea behind this proposal is that during the convergence process, the perceptual distance between this image and the following ones ($I^{(p)}$ with $p > 1$) increases. As a consequence, the perceptual distance between $I^{(p)}$ and $I^{(p+1)}$ decreases when p increases. Thus the value $VDP(I^{(1)}, I^{(p)})$ converges to a maximum and the VDP distance curve will highlights an horizontal part for any value greater than the perceptual threshold.

We used a linear regression to estimate the director coefficient of the line which passes by the slop and to predict the function linearity. For this purpose we keep in memory the last N VDP values computed before the current image. They are used for recalculating the new regression coefficient and rendering is stopped when it becomes less than a predefined small value ε.

4.2   Results

Our approach and the one of Myszkowski have been implemented and tested on the 8 images of figure 1. The results are presented in table 2. It appears that our results

Table 2 The results provided by our approach and the one in [Mys98] as compared to the experimental sampling thresholds

Scenes	number of light samples/pixel		
	thresholds	our approach	[Mys98]
BAR 1	3104	3000	5400
BAR 2	3808	3900	7200
CUBE 1	2157	2200	4200
CUBE 2	3053	3000	6200
OCULIST 1	3270	3300	6200
OCULIST 2	4322	4400	8000
PLANTS 1	2667	2600	4400
PLANTS 2	2701	2800	4800

are very closed to the experimental thresholds. These ones correspond to 90% of the subjects that do not perceive any difference between the reference image and the image computed with the number of light samples that appears in the second column of this table.

As previously introduced the relation proposed in [Mys98] overestimates the thresholds: our approach performes about 40% faster by using more rigorous features.

4.3   Advantages and Drawbacks

The VDP is a complex model which purpose is to reproduce the general features of the human visual system. Its goal is to allow to detect any kind of difference between two images such as noise, geometry, intensity or colors. We are interested only in noise detection; even if our results appear to be interesting, a more specific approach could be more useful and accurate than the VDP, specifically by considering the following questions.

Indeed the HVS is intricate and composite. Therefore the majority of the models that have been applied to computer graphics have been simplified and/or modified as compared to the original ones and to the "reality".

By considering the fact that the value of some parameters of Daly's VDP are unkwown, Myszkowski [Mys98] had to initialize and calibrate them in order to be usable in global illumination methods. Yee [Yee04] proposed an abridged VDP version in which he removed some of the more expensive computations of Daly's algorithm and replaced them with approximations.

These models were originally validated by neuro-biological and psycho-physical studies. But not all of their simplifications or modifications have then been validated. For example, Chalmers et al.[LC04] have shown through experimental results that the VDP does not always give accurate responses. Takouachet [Tak09] studied the use of Yee's VDP and noted that its results are often far from those of the original method.

Finally the VDP is used on full images. Even if global perceptual thresholds are helpful in allowing to stop stochastic global illumination, they do not allow to consider heterogeneity of noise through any image.

5 SVM

In this section we investigate the use of classification methods that are often used in image analysis. Indeed our final goal is to develop a software component that should be able to mimic the human visual system and its capability to detect noise in any image. It means that this component should be able to classify any part of an image as noisy or not without having recourse to any HVS model.

Support vector machines [Vap95a, Vap95b] are part of a set of supervised learning methods for regression and classification problems. SVMs compute an hyperplane that provides an optimal separation of data. Linear SVMs are known to be maximum margin classifiers and to minimize the classification error. A major property of SVMs is their ability to work with large dimensional problems and to find complex separation planes: if the problem is not separable in the current space, data can be projected in larger spaces where separation could be easier.

The advantages of such approaches is that they rely on real cases, meaning that learning is performed directly through the use of experimental data. Then they have been shown to be robust which is of great interest in our case when the images to be analyzed are not part of the learning images set. Finally their use is fast once learning has been done.

However one must take care in the data that are used during the learning stage of the method since it has a great impact on what is learned and thus how classification will be then performed.

5.1 Overview of the Approach

SVMs use supervised learning; that means that their use is performed in two steps (see figure 4) :

- the first one focuses on the learning part of the method. Some examples must be provided to the method with an information about the class to which each data belongs to;
- the second step uses the learning database for classifying data that have not been learnt previously by the method.

During the first step we thus have to provide the SVMs with some images and a flag specifying whether each image is noisy or not. Once the supervised learning step has been performed the corresponding database can be questioned in order to know whether any other image is noisy or not.

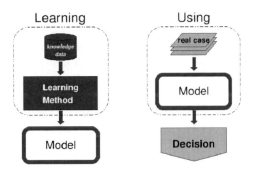

Fig. 4 The two steps of a learning approach: training the method from some examples and then using the model for taking decision

5.2   The Learning Stage

In sections 3.1 and 3.2 noise thresholds have been obtained from experimentations with human subjects. We chose to use the data obtained for the subimages for two reasons: performing supervised learning on small subimages will allow to take into account noise heterogeneity which is of great interest for stopping calculations independently for each subimage; then it will increase the efficiency of the learning-step by reducing the number of pixels that have to be associated to each learning example.

5.2.1   The Image Dataset

Training efficiently any learning method requires first to have a large set of examples. In our case we computed a view of 12 virtual scenes, meaning we got 12 images of size 512×512. By cutting each of those images in 16 subimages of size 128×128, we got a total of $12 \times 16 = 192$ different subimages. Each of this subimage has been computed with a different number of light samples, from 100 to 10.000 samples per pixel by step of 100 samples. Thus we got 100 versions of each subimage, that is 192×100 subimages with different content and noise levels.

Thanks to the experimentations described in section 3.2, we got the perceptual noise threshold for any of the 192 converged subimages. The set of all the subimages is then cut in two parts: one half for the training of the SVM method and the second one for validation purpose.

5.2.2   The Training Protocol

In the training protocol we experimented, pairs of subimages are provided to the method: a subimage S_n^{ref} so called the *reference subimage* and one of the test subimages $S_n^{(p)}$. A third information is combined to the subimages, that states whether the two subimages are considered as identical or not: the reference subimage and the test one are stated as identical when the index p of the test subimage is greater than the noise threshold. All pairs of subimages $(S_n^{ref}, S_n^{(p)})$ of the training set are thus randomly provided to the learning method. According to the size of the training set, we were able to provide 9.600 pairs of sub-images to the learning model. This should ensure a sufficient training data set since SVMs are known to be efficient even on small sized example sets.

However the same problem as for the VDP approach arises: ideally the reference subimage should be the converged one and is still unavailable. We proposed to replace this subimage by one that must be noisyless and that should highlights the same main visual features than the converged one. These important features are shadows, reflections, colours and textures. Even if all the lighting features that are computed by global illumination are not present in this image it is expected that it is sufficiently closed to the converged one for allowing the trainign step to be efficient. Different techniques could be used for computing this image like OpenGL rendering including shadows and reflections or ray tracing. We chose to use a rendering using ray tracing. Thus for each image to be used in the training protocol and further for the use on the model database, a ray traced image is computed with the same point of view and size as the noisy ones. Similarly it is cut into sixteen subimages, each one being assumed to be the reference subimage. Then during training each noisy or noisyless subimage is provided to the SVM classifier with this reference image and the information about their difference (see figure 5).

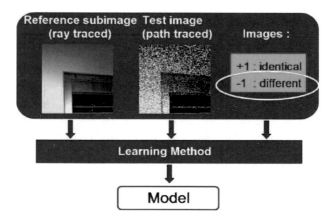

Fig. 5 The training protocol: a ray traced reference subimage and an example subimage are provided to the learning method with information about the fact that the two subimages are identical or not from the point of view of noise

Note that ray tracing the reference image is performed once both for training and using the model and that the corresponding time is generally very low (some seconds) as compared to the computation times of the Path Tracing method. However it is obvious that some lighting effects cannot be taken into account by ray tracing (*i.e.* high indirect lighting, caustics). In these cases the method we describe below fails to correctly classify noisy and noisless subimages due to a bad reference subimage.

5.2.3 The SVM^{light} Library

In this paper we used the free SVM^{light} library[2] for all our experimentations. This library is an implementation of Vapnik's Support Vector Machine [Vap95a] for the problems of pattern recognition, regression, classification and for the problem of learning a ranking function. The optimization algorithms used in SVM^{light} are described in [Joa00].

SVMs can be parametrized with different kernels among them linear and polynomial kernels or radial basis functions. These kernels are used for projecting data into multidimensional spaces. Radial Basis Function (RBF) kernels are widely used for the reduced set of parameters that have to be tuned and then because they provide robust learning models for lot of non linear classification problems [CHT02][MB04]. Nonetheless we studied in [Tak09] the use of several of those kernels and showed that RBF kernels provide the best results on our classification problems.

5.2.4 The Data Format

Subimages format has to be modified in order to be used for the training step. First the RVB subimages issued from the path tracing method are converted into the YC_rCb colour model and only the luminance is saved separately as a vector of numerical values (noise is mainly characterized by luminances rather than chromaticities).

This vector and its corresponding reference vector can then be submitted to the SVM learning process. However we shown experimentally [Tak09] that this simple mapping (colored pixels to luminances vector) is inefficient and the results of learning are often of poor quality. The images contain more information than only noise and this information appears to be used during the classification process. We thus propose to highlight the high frequencies noise by using a method known as the *noise mask*. This denoising technique is commonly used in satellite imagery [Rus92]. It uses a blurr mask for enhancing details and reducing the images noise. It is applied in two steps: firstly a blurred image is computed using a 3×3 gaussian convolution using a convolution coefficient $\sigma in [0.3, 1.5]$. Then the noise mask is obtained by computing the difference between the original image and the blurred one.

The noise is then reduced by subtracting α times the noise mask from the original image:

$$Image_{new} = Image - \alpha \times Noisemask$$

We are not interested here in denoising the images but rather in locating the areas where noise affects the image. Thus we experimented the use of the noise mask as the converting tool for the images that are submitted to the learning process. Using such an approach leads to a classification accuracy of 98% with a high number of support vectors which is a good indicator of the learning efficiency of the model.

---

[2] http://svmlight.joachims.org/

5.3   The Algorithm

Once learning has been performed we are able to use the corresponding classification model for noise detection. We applied it in a progressive path tracing algorithm where $N = 100$ new light samples are added at each iteration to the unconverged subimages.

Then the model is asked whether each new sampled subimage is still noisy or not. According to the model's answer we then decide to add new samples or to stop computation for the corresponding subimage as it is supposed to be visually converged. Like for the learning step the model is interrogated by furnishing it both the currently computed subimage and the reference one which has been computed once with a ray tracing method. Note that questioning the model is very fast since it requires only a few milliseconds per subimage.

Table 3 The average number of samples per pixel required for each scene to be perceived as not noisy (exp.: experimental values; model: values obtained with the learning model)

scene	experimental threshold	model threshold	scene	experimental threshold	model threshold
Occulist	3278	3287	Deskroom1	3030	3012
Cornell box	2344	2300	DeskRoom2	2481	2581
Taproom1	3234	3181	Taproom2	2816	2893
Bakery	2215	2212	Classroom	2255	2300
Ironmonger	2385	2381	Draper	2767	2737
Sponza	2900	2862	Grocery	3168	2968

3474 / 3400	2568 / 2500	2219 / 2300	2114 / 2000
3171 / 3200	2590 / 2600	3077 / 3200	3123 / 3000
3144 / 3100	3039 / 3000	2423 / 2500	2474 / 2500
2624 / 2700	3117 / 3000	2718 / 2900	2399 / 2400

Fig. 6 The results of noise detection on one of our scene. (left) the 16 subimages ; (right) the table indicates the experimental thresholds (up) and those generated by the model (bottom)

Fig. 7 The 12 images used during the experiments with the SVMs. The images in the first two rows have been used for the learning step ; the ones in the two last rows have been used for validation purpose only

5.4   Results

We compare in table 3 average experimental thresholds for 12 test images (see figure 7) with those provided by the classification model. It does not give details subimage per subimage but these results highlight the compatibility of both values. Note that the left scenes were used for the learning stage and the right scenes were never learnt by the model.

By analysing more accurately the results for each subimage, we note that both threholds are always very closed. An example of such analysis is provided in figure 6. The subimages of this image (scene I11 in table 3) were not learned by the model. The first number indicated for each sub-image represents the average number of light samples per pixel required for 95% of the subjects to see the subimage as not noisy. The second number is the number of samples computed by the model; it thus represents the stopping threshold for the iterative path tracing algorithm.

6   Conclusion and Perspectives

We presented two methods that allow to detect the visual convergence of global illumination algorithms, more specifically Path Tracing based methods. The first one relies on the use of the VDP perceptual model and is able to find an image global convergence threshold that is closed to human subjects data. The use of the VDP for detecting convergence more locally on subimages has to be studied carefully since it would reduce considerably the computation times. The second approach we detailed attempts to avoid the use of any complex perceptual model which are often incomplete and difficult to parameterize. It relies on supervised learning of noise detection and provide good local results and fast classification. But its current version depends on the use of a reference image that is computed by the ray tracing method. Even if this method provides generally fast results, it lacks in capturing complex lighting effects and thus could provide bad learning models. This problem has to be solved in order to be able to provide an efficient tool for improving the use of photorealisitic rendering softwares.

References

[CHT02]   Davis, L.S., Huang, C., Townshend, J.R.G.: An assessment of support vector machines for land cover classification. International Journal of Remote Sensing 23, 725–749 (2002)

[Dal93]   Daly, S.: The visible differences predictor: an algorithm for the assessment of image fidelity. In: Digital Images and Human Vision, vol. 4, pp. 124–125 (1993)

[FP04]   Farrugia, J.-P., Péroche, B.: A progressive rendering algorithm using an adaptive perceptually based image metric. Comput. Graph. Forum 23(3), 605–614 (2004)

[Itt00]   Itti, L.: Models of Bottom-Up and Top-Down Visual Attention. bu—td—mod—psy—cv, Pasadena, California (January 2000)

[Joa00]   Joachims, T.: Estimating the generalization performance of a SVM efficiently. In: Langley, P. (ed.) Proceedings of ICML 2000, 17th International Conference on Machine Learning, Stanford, US, pp. 431–438. Morgan Kaufmann Publishers, San Francisco (2000)

[Kaj86]   Kajiya, J.T.: The rendering equation. SIGGRAPH Comput. Graph. 20(4), 143–150 (1986)

[KU85]   Koch, C., Ullman, S.: Shifts in selective visual attention: Towards the underlying neural circuitry. Human Neurobiology 4, 219–227 (1985)

[LC04]   Longhurst, P., Chalmers, A.: User validation of image quality assessment algo-
            rithms. In: Theory and Practice of Computer Graphics, EGUK 2004, pp. 196–202.
            IEEE Computer Society, Los Alamitos (2004)
[Lux]   LuxRender, http://www.luxrender.net/en_GB/index
[MB04]   Melgani, F., Bruzzone, L.: Classification of Hyperspectral Remote Sensing Im-
            ages With Support Vector Machines. IEEE Transactions on Geoscience and Re-
            mote Sensing 42, 1778–1790 (2004)
[Mys98]   Myszkowski, K.: The visible differences predictor: applications to global illumi-
            nation problems. In: Eurographics Rendering Workshop, pp. 233–236 (1998)
[Nex]   NextLimit, http://www.maxwellrender.com/
[NIK01]   Niebur, E., Itti, L., Koch, C.: Controlling the focus of visual selective attention.
            In: Van Hemmen, L., Domany, E., Cowan, J. (eds.) Models of Neural Networks
            IV. Springer, Heidelberg (2001)
[PFFG98]   Pattanaik, S.N., Ferwerda, J.A., Fairchild, M.D., Greenberg, D.P.: A multiscale
            model of adaptation and spatial vision for realistic image display. Computer
            Graphics 32(Annual Conference Series), 287–298 (1998)
[RPG99]   Ramasubramanian, M., Pattanaik, S.N., Greenberg, D.P.: A perceptually based
            physical error metric for realistic image synthesis. In: Rockwood, A. (ed.) Sig-
            graph 1999, Computer Graphics Proceedings, Los Angeles, pp. 73–82. Addison
            Wesley Longman, Amsterdam (1999)
[Rus92]   Russ, J.C.: The Image Processing Handbook. CRC Press, Boca Raton (1992)
[Sar97]   Sarnoff Corporation. Sarnoff JND vision model algorithm description and testing,
            VQEG (August 1997)
[Tak09]   Takouachet, N.: Utilisation de critères perceptifs pour la détermination d'une con-
            dition d'arrêt dans les méthodes d'illumination gobale. PhD thesis, Université du
            Littoral Côte d'Opale (January 2009)
[TDR07]   Takouachet, N., Delepoulle, S., Renaud, C.: A perceptual stopping condition
            for global illumination computations. In: Proc. Spring Conference on Computer
            Graphics (2007)
[TG80]   Treisman, A.M., Gelade, G.: A feature-integration theory of attention. Cognit.
            Psychol. 12(1), 97–136 (1980)
[Vap95a]   Vapnik, V.: The Nature of Statistical Learning Theory. Springer, New York (1995)
[Vap95b]   Vapnik, V.: Statistical Learning Theory. John Wiley and Sons, Inc., New York
            (1995)
[VG97]   Veach, E., Guibas, L.J.: Metropolis light transport. Computer Graphics 31(Annual
            Conference Series), 65–76 (1997)
[Yee04]   Yee, H.: A perceptual metric for production testing. Journal of Graphics
            Tools 9(4), 33–40 (2004)

Verbalization of 3D Scenes Based on Natural Language Generation Techniques

Vassilios Golfinopoulos[1], Dimitrios Makris[1], Georgios Bardis[1],
Georgios Miaoulis[1,2], and Dimitri Plemenos[2]

[1] Technological Education Institution of Athens
   Department of Informatics
   Ag.Spyridonos St., 122 10 Egaleo, Greece
   golfinopoulos@cs.teiath.gr, demak@teiath.gr,
   gbardis@teiath.gr, gmiaoul@teiath.gr
[2] XLIM Laboratory, University of Limoges,
   83 rue d' Isle, Limoges, 87000, France

Abstract. Basic ideas and requirements can be outlined by a design through ambiguous terms in order to define a desirable scene. Declarative modeling approach receives a rudimentary description and produces a set of scenes that are close to designer view. The reverse declarative modeling paradigm supports the designer to distinguish a set of scenes, accommodate further the pre-selected scenes to his needs, and produces a new enriched declarative description which initiates a new forward declarative design cycle for new promising scenes. The aim of the present work is to enhance the communication between the designer and machine, in such a way to increase the designer understanding and perception, by structuring a description in textual mode, reflecting all necessary semantic and geometric information, whenever the designer alters the pre-selected scenes.

Keywords: Natural Language Generation, Communication, Scene Understanding, Declarative Modeling, Semantic Model, Knowledge-Based Systems, Reverse Engineering.

1 Introduction

Communication is based upon the exchange of any kind of signs. A characteristic communication episode consists of seven processes, taking into account the existence of a sender and a receiver. In any mode of communication the process involves firstly, the intention, generation, synthesis and secondly, the perception, analysis, disambiguation, incorporation on the other hand [4]. The intention focuses on the intention of the speaker to announce something that is worth saying. The generation focus upon the creation of an utterance that includes the meaning and in the synthesis stage, the speaker produces the realization of the utterance via some medium.

D. Plemenos and G. Miaoulis (Eds.): Intelligent Comp. Graphics 2011, SCI 374, pp. 19–39.
springerlink.com   © Springer-Verlag Berlin Heidelberg 2012

Declarative modeling [1] is an alternative modeling paradigm that adapts the design process, overcomes the disadvantages of geometric modeling and allows the designer to describe the desired scene by defining its properties, which can be either precise or imprecise, and without indicating the way to obtain a scene with these properties. Declarative modeling liberates the designer from defining the geometric properties of the entities and facilitates the designer in order to describe the scene by requiring only some already known properties. A special approach of the declarative modeling is the declarative modeling by hierarchical decomposition [1, 2]. This approach provides the designer describes the desired scene by top-down decomposing the scene at different levels of details and facilitates the description of complex scenes.

The declarative scene modeling is based on the declarative design cycle, which consists of three sequential functional phases [2], (Fig. 1). The first is the Scene Description phase, where the designer describes how he perceives the scene by specifying properties of the scene or leaving them ambiguous. The second is the Generation phase, where the generator inputs the declarative model and produces a set of solutions that meet the description of the desired scene. The third is the Solution Understanding phase, where the scene solutions are visualized through a geometric modeler.

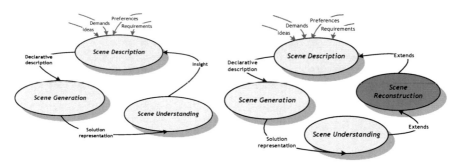

Fig. 1 Declarative modeling process   **Fig. 2** Reverse Declarative modeling process

1.1 Declarative Reverse Design Methodology

The Declarative Reverse Design is a novel approach [13], which constructs a new declarative description from an initial selected three-dimensional geometric solution, which has been modified by the designer, in order to enable the design process to become iterative automatically until the system produces the most desirable solutions. The declarative design methodology starts with the description of the desired scene in terms of objects, relations and properties. A rule set and an object set are built representing the designer requirements of the scene.

Initially, the object set consists of all objects of different level of detail, and the rule set consists of all relations, properties that the designer has declared during the declarative description phase. Based on that rule set, the solution generator, which uses constraint satisfaction techniques, produces a set of scenes, which are visualized and the designer selects the most desirable solutions, which can be edited.

The declarative design cycle can be extended to an iterative process by using a Reconstruction phase [13] where the selected scenes are understood semantically and refined by adding more detailed descriptions in successive rounds and leading to more promising scenes, (Fig. 2). Under the Reconstruction phase, the designer changes the geometry of the selected scenes by modifying the topological relations and geometric aspects of the objects. These changes are checked semantically and the special representations are updated. The Reconstruction phase carries out the declarative reverse design methodology, receives a set of selected geometric models and converts each of them into a stratified representation. The system constructs for every selected scene a stratified representation. The rule set and the object set of each scene can be edited by adding, deleting, and changing the objects, relations and properties of the scene. The designer can proclaim his/her requirements declaratively and geometrically during the reconstruction phase. A new declarative description is constructed by the semantic model, which is constructed by unifying all already modified stratified representation along with the rule and object set.

1.2 Research Scope

The current work aims to provide an enhanced reverse declarative design approach by exploiting methods and approaches from the Natural Language Generation field. Within the broad context of the synthesis of three-dimensional scenes with the aid of reverse declarative modeling, the research objective is to enable a semantic enrichment of the declarative description within the reverse design processes. In this way, the paper objective is the definition and implementation of a natural language generation component. The aim of the present work is to enhance the communication between the designer and machine, in such a way to increase the designer understanding and perception, by structuring a description in textual mode, reflecting all necessary semantic and geometric information, whenever the designer alters the pre-selected scenes

Before the detailed discussion of the proposed approach, we first briefly present some relevant work, and second we justify the necessary Natural Language Generation search principles in our approach.

2 Relevant Work

In this part, we will present some works and current developments within the overlapping areas of reverse declarative scene modeling, and natural language generation.

2.1 Reverse Declarative Scene Modeling

Numerous approaches have been developed in order to aid the declarative scene modeling – design, and within a variety of application domains (computer graphics, virtual reality, architecture, urbanism, product design, et cetera). In particular, all approaches have been targeting to facilitate users during the solution generation and understanding phase in order to tackle the numerous generated alternatives. However, only one research work has presented an early methodology towards a reverse engineer framework, in order to convert a geometric description to a declarative one. The XMultiFormes project [30] tries to integrate the two modelers by using a special interface system to ensure that there is full and complete transfer of information between the declarative and a traditional geometric modeler. This system is composed of four sub-processes, each of which is responsible for one aspect of the information transfer. The geometric convection process translates the geometric representation to one that is more suited to interactive modeling. The labeling system is responsible for capturing non-geometric information, which is implied in the declarative description. The geometric-to-declarative representation conversion process converts a geometric instance to declarative description and the scene inclusion process provides a means for the inclusion of previously generated scenes in a declarative description. The XMulti-Formes project does not incorporate a knowledge based management system since it gives special emphasis on retrieving the appropriate knowledge for reasoning from the designer.

2.2 Natural Language Generation

A great number of works in Natural Generation area has focused on introducing implementation approaches for Natural Language Generation, (henceforth NLG) system architecture. In Reiter, E., Dale, R [15] it is described a generic model of NLG system architecture. A generic architecture for generation is presented in Mellish C., Evans R. [16] where a distinction is set between the functional and implementation aspects of the architecture. The functional architecture refers to the data representations and the definition of modules while the implementation architecture refers to the software services. The work of Mellish, C. et al. [20] provides a reference architecture that is based on data models for key knowledge resources and in Paris C. et al the presented work [19] addresses a platform focuses on reusable mechanisms for content modeling and discourse planning.

Many applications-specific NLG systems have been developed that refers to the generation of dynamic hypertexts [21], instructional texts [22], and navigation instructions in a virtual environment by using Artificial Intelligent planning- based approach [18]. An effective method for generating natural language sentences [17] is based on a hybrid tree representation that encodes the meaning representation along with the natural language. The phrase-level dependencies are modeled among language phrases and meaning representations, by using a tree conditional random field on top of the hybrid tree. The study of Fu Ren, Qingyum Du [27] presents the transformation of kinds of maps and spatial orientation information into natural language text.

NLG applications differs from the input to the system, e.g. the work of Kosseim, L., and Lapalme, G. [23] uses traces of tic-tac-toe games, in Goldberg, E et al [24] uses data from weather reporting workstations, in Karberis G., Kouroupetroglou G. [28] uses spontaneous telegraphic and generates well-formed sentences for the Greek language. The work of Lavoie, B et al [25] faces object-oriented software models. In Atsuhiro Kojima et al [29] video images are transformed into textual descriptions by applying by semantic primitives and [26] handles task models.

3 RS-MultiCAD System Architecture

MultiCAD architecture [5] is an intelligent multimedia Computer-Aided-Design system. The design environment of MultiCAD features a rich set of modules. These include alternative modules for solution generation using constraint satisfaction programming [6] or genetic algorithms [7] as well as modules responsible for introducing architectural knowledge [8], representation of architectural styles [7], collaborative design [9], and intelligent user profile [10]. MultiCAD incorporates an object-relational database [12], which consists of five logical interconnected databases. The scene database is supporting information describing the scene models. The multimedia database is containing all types of documents related to the project. The knowledge base is containing all the necessary information about type of objects, their properties along with their relations. The project database is manipulating with data concerning planning, financial and other special aspects of each project and finally the concept database [11] is storing concepts representations. The scene database is configured following the Scene Conceptual Modeling Framework [12]. The description contains objects defined by their properties, simple or generic ones, as well as group of simple objects with properties in common. Besides, the description contains three types of relations between objects:

- Meronymic (*"is part of"*, *"consists of"*),
- Spatial organization ("adjacent south", *"equal length"*) and
- Reflective ("higher that large", *"wider than deep"*) relations.

Finally, the description also contains properties which describe objects (is long = '*Low*').

RS-MultiCAD is a knowledge-based system that implements the declarative reverse design. The RS-MultiCAD knowledge-based component incorporates architectural domain specific knowledge for constructing buildings (Fig. 3). The basic system architecture is modular and consists of five main modules.

The Import / Export module is responsible for the communication with the databases supporting the scene input and output, the output of a modified declarative description and finally the Import / Export of geometrical models of other types.

The Extraction module applies all domain specific relation and property types and extracts all valid relations, properties of the objects from a selected solution. The Extraction module is domain independent and facilitates the extension of knowledge since it parses the available knowledge from the database.

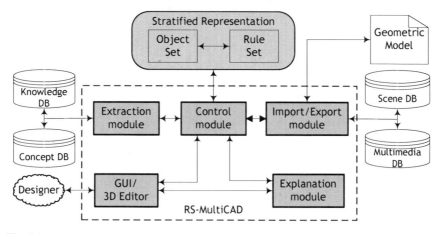

Fig. 3 Basic system architecture

The Control module incorporates all necessary mechanisms for building, manipulating, updating the stratified representations and unifying them into semantic model. The stratified representation is dynamic and constructed from the designer selected solution with a top-down approach and mainly consists of declarative and geometric information. Declarative information can be summarized into object set and rule set. Geometric information deals with the geometry of each object that constitutes the scene. The control mechanism is event-driven and is responsible for the stratified representation to ensure the correct transition from one state to another. It handles the designer scene modifications examining their semantic correctness and properly updates the stratified representation by propagating the changes in a mixed way.

The explanation module provides valuable information about the system reasoning in cases where a scene modification violates the rule set. Finally, the RS-MultiCAD system incorporates a graphical user interface with a 3D editor in order to visualize the solutions and graphically receive the designer requests.

From another point of view, the designer selects a set of scenes which has been produced upon his/her initial requirements. The system (Fig. 4) constructs a set of stratified representations and properties, relations are extracted based on the geometry of every scene. Every scene has its own partial stratified representation comprises of a rule and object set. The designer modifies the scenes, these changes are checked for their validity, and if so, the system propagates these changes by updating the respective partial stratified representations.

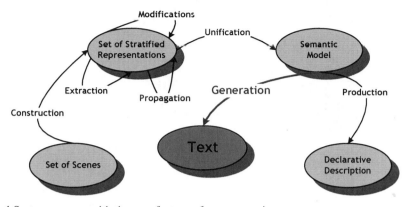

Fig. 4 System process with the new feature of text generation

The unification of the partial stratified representations leads to a semantic model. The construction of the semantic model is accomplished by unified all partial rule sets. Concerning the spatial organization relations the system unifies complementary spatial relations between two specific objects by adding to the unified rule set of the semantic model, the disjunction of the initiatory relations. Concerning the reflective relations, the system unifies complementary reflective relations which refer to the same object I different scenes by adding to the unified rule set of the semantic model, the disjunction of the relations. The unification of object properties, which refer to the same object in different scenes in terms of different property values, is implemented by placing to the unified rule set of the semantic model, the disjunction of the object property initial values.

It must be pointed out that the role of the semantic model is crucial since the system based upon it, produces the text along with the new declarative description, supporting the iterative nature of the declarative conception cycle.

3.1 The Stratified Representation

The need of representing geometrical and declarative information leads to an approach of using a stratified representation [14]. A model in order to become another type of model is gradually transformed into a sequence of different levels of abstraction by a sequence of processing steps.

The stratified representation is an intermediate level model necessary for connecting the declarative with the geometric model, and embodies the two distinct interconnected layers of representation, the declarative layer that represents the scene description with the hierarchical decomposition, and the geometric layer that encapsulates the geometric aspects of the objects, (Fig. 5).

The geometric layer of the stratified representation is based on the bounding box dimensions of each object, which express the object pure geometric properties, along with any extra geometric information that can determine the shape of the object.

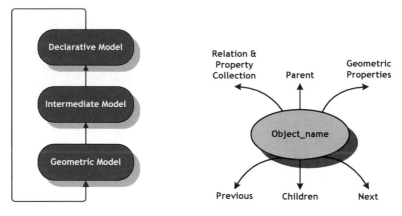

Fig. 5 Declarative modeling design   **Fig. 6** Internal object structure
process

RS-MultiCAD inputs a set of geometric models produced by the solution generator. That geometric model contains the geometric information of all objects and their type as well. The stratified representation is a dynamic semantic net with nodes and directed arrows. Every node corresponds to an object that participates in a scene, (Fig. 6). The arrow label indicates the relations of the nodes. The labels "*Parent*" and "*Children*" connect nodes with same level of abstraction and represent the meronymic relations. The labels "*Next*" and "*Previous*" connect nodes with the same level of abstraction and detail. The label "*has-geometry*" connects nodes of different layers and represents the geometry of an object. Finally, the label "*has-topology*" connects nodes of the same level of abstraction indicating the topological relations among concepts and represents the reflective and spatial relations. Fig. 6 presents the basic structure of the object.

The construction of the stratified representation is a top-down process where the hierarchical decomposition is built based on the geometric information coming from the geometric model. For every object, a node is created on the geometric layer of the stratified representation. As long as all nodes have been created, the pure geometric properties lead to the hierarchical decomposition by creating interconnected nodes on the declarative layer of the representation.

3.2 Altering the Scenes

RS-MultiCAD allows the designer to perform geometric and topological modifications on the scene independently. As soon as the designer modifies the scene, a special process starts. Every designer's modification must be checked according to the rule set for its validity and if so the stratified representation must be properly updated in order to reflect the real state of the scene. RS-MultiCAD provides two inference options according to designer modification, which may or may not be activated.

The first option refers to check the modification according to the rule set. A modification is valid as long as no relation or property of the rule set is violated otherwise the modification is invalid and is canceled. If the designer decides not to check the modifications according to the rule set, the control module performs a set of mandatory conditions ensuring the validity of the scene such as, non-overlapping objects of the same level of abstraction, no object exceeding the over-all scene limits, etc.

The second option refers to add pure geometric properties to the rule set that are inferred from the modifications. If the designer moves an object to a new posi-tion, pure geometric properties relative to move are added in the rule set.

The control module properly propagates the modification by updating the geo-metric layer of the representation and activating the extraction module in order to recalculate all valid relations and properties. If none relations and properties of the rule set are violated the changes are accepted and the new state of the stratified re-presentation is valid. Otherwise, the explanation module is activated in order to record all violated relations, properties of the rule set and the control mechanism rolls the representation back to the previous state.

The modifications that can occur on the stratified model they refer to an ab-stract or leaf node and can be divided into two categories according to the existed geometrical information that may be supplied by the designer, first the declarative modifications and second the geometrical modifications. In particular, the declara-tive modifications are the following:

- The insertion of an abstract node in the stratified model can be done by specify-ing firstly an already existing node of the model as its parent and secondly the nodes that become children of the new abstract node. The result of such a change will affect the stratified representation since the object set changes.
- The deletion of an abstract node will eliminate the sub-tree where the abstract node is root. The result of such a change will affect the object set and may af-fect the rule set as well. The stratified representation must be updated in order to reflect the current state of the scene.
- The designer changes the rule set by adding or deleting a relation or a property of a node.

Moreover, the geometrical modifications on the respective layer refer to move, scale, resize, insert of an object. Additionally such modifications may alter any ex-tra geometric characteristic of an object. Every time a modification on a specific scene occurs, the system updates the specific stratified representation, object and rule set and rebuilt the semantic model [3] along with the other scenes that belong to the set of the pre-selected scenes, since every selected scene has its own object set and rule set.

3.3 NLG Component

Generally speaking, it is specified [15] that the input of an NLG system can be considered as a four-tuple vector (k, c, u, d), where:

- k stands for the knowledge sources that are used by the system,
- c stands for the communicative goal to be achieved,
- u stands for the user model, and
- d is a discourse history.

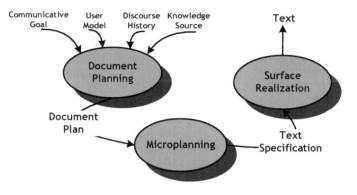

Fig. 7 NLG typical architecture

The typical architecture of a Natural Language Generation system consists of three main phases:

- the document planning,
- the microplaning and
- the surface realization.

The typical architecture of a NLG system is presented in Fig. 7. Especially, the knowledge source can be all relevant information about a specific domain, gathered in typical knowledge bases. The communication goal defines what one intends to express and what it is for. The proposed system has a specific aim to provide all the information imposed by the semantic model.

Furthermore, the user model describes the reader for whom the text is to be generated. According to the user model, the NLG system should configurate the content of the text to be generated upon the reader. The proposed system intends to produce standardized content texts that are read by designers-architects.

The discourse history describes what information has been said to the reader so far in order to avoid repetitions, already mentioned entities and properties. In the proposed system, the new placed information, according to previous text, is displayed in bold font style.

During the document planning phase (Fig. 8) the system, makes decisions of what information should be transferred to the reader and how the final output, the text should be structured. The document-planning phase incorporates two main stages [15]. The content determination refers to the choice of what information must be placed on the output text. The second stage, the document structuring refers to the ordering and the structure of the output text.

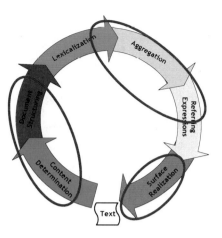

Fig. 8 NLG typical architecture

In our proposed system, the input of the NLG component is the semantic model, namely the unified stratified representation along with the object set and the rule set.

The relations and properties, which have been selected by the designer-architect, form the messages. The extraction of the messages is implemented using a top-down approach. The system supports two types of messages (a) and (b):

$$[\ X \ \ R\_P \ \ Y \ ] \quad \text{(a)}$$

where X signifies an object name, R_P signifies for a relation or property and Y signifies an object name or value of property R_P or null, for example:

[Bed3 adjacent_under Roof5]
[Bed3 higher_than_large]
[Bed3 length 3]

$$[\ X \ \ Y \ ] \quad \text{(b)}$$

where X signifies an object name and Y signifies the class name of the object X, for example.

[B4 Bedroom]

The generated text is planned to consist of a number of paragraphs. The number of paragraphs depends on the number of objects that contribute to the scene. Every paragraph refers to an object and the system places firstly the message, which refers to the type of the object, secondly the messages that refer to the relations, and finally the rest messages, which refer to the properties of the object.

During the microplanning phase (Fig. 8) the system selects the appropriate words, such as verbs nouns adjectives and adverbs, and syntactic constructions in order to convey the meaning of the content selected by the previous phase. The microplanning phase incorporates the lexicalization, aggregation and referring expression generation stage.

The lexicalization stage (Fig. 8) focuses on choosing the appropriate words and syntactic structure that express each message. In our case, the lexicalization is based on templates, pre-defined phrase specifications. For each message, the system produces an abstract sentence description, with abstract object names and ignores the grammar correctness temporarily since this is realized in the surface realization phase. Each message is associated with at least one template, which contains a number of slots and information about syntactic terms. For example the messages indicating that a site consists of a building and a garage are:

[Site consist Building2]
[Site consist Garage3]

The messages are associated with a specific relation template and the system has to fill in the slots:

[Ref_Exp (nominative, "%Site"), Verb (active, present, third, "consist"),
  Preposition ("of"), Ref_Exp (accusative, "%Building2")]

[Ref_Exp (nominative, "%Site"), Verb (active, present, third, "consist"),
  Preposition ("of"), Ref_Exp (accusative, "%Garage3")]

The knowledge source contains lexicons containing all necessary information about noun, noun phrase such as the gender (nonpersonal), number (singular, plural), countable or uncountable, verb such as the voice (active, passive), person, number (singular, plural) along with prepositions and articles.

The output of the lexicalization is a set of simple, short abstract sentence specifications. The stage of aggregation is responsible for sentence formation by combining multiple sentence specifications into one sentence by applying specific operations [15].

The simple conjunction operation combines two or more sentences within a single one by placing a conjunctive such as the "*and*". The conjunction via shared participants operation [15] is applicable if there are two or more sentences which share the same syntactic elements then it is possible to combine them by appearing the common elements only once. The operation is applicable since two sentences share the same subject and verb or verb and object. The operation is applied to the previous example resulting the following:

[Ref_Exp (nominative, "%Site"), Verb (active, present, third, "consist"),
  Preposition ("of"), Ref_Exp (accusative, "%Building2"),
  Conjunctive ("and"), Ref_Exp (accusative, "%Garage3")]

The syntactic embedding operation refers to the hypotactic aggregation [15]. The operation is applicable in our case if there are two sentences referring to an object property numerically and abstractly. The below example shows the effect of the specific aggregation

During the referring expression generation stage (Fig. 8), the abstract elements are replaced by the semantic content of noun phrase referring expressions. In the proposed system, every time the system starts describing an object, it uses its real name for the very first time and thereafter an appropriate pronoun such as "*it*", is used whenever the system refers to that object.

The system, through the surface realization phase, converts the abstract sentence descriptions to text. Such conversion is realized by replacing the abstract object names with real names and placing punctuation marks. Besides, the system applies syntactic and grammatical rules concerning the appropriate order of the syntactic terms, the placement of prepositions and articles, the agreement of the case and singular or plural form between subject and verb and the agreement that the object complement must be placed in the accusative case.

3.4 Solution Generalization

The system returns the semantic model required by MultiCAD in order to reduce, in the next design phase, the initial solution space and generate solutions that are more promising. The RS-MultiCAD provides two optional ways, the manual and automated. In particular, RS-MultiCAD in the manual way results in a semantic model that is based on the initial rule set along with the new relations and properties that have been changed by the designer. In this way, RS-MultiCAD offers the designer the possibility to drive the system to generate a solution space that is nearer to his requirements.

Furthermore, the automated way is based on the generalization factor. Every hierarchical decomposed tree is divided in distinct levels of detail. The generalization factor is related to levels of detail, and its values vary from one to maximum tree depth. The semantic model that results from the automated option is based on the initial rule set along with all modifications and also all pure geometric properties that are implied from the generalization factor.

4 Case Study

The case study framework aims to address the feasibility of the proposed and implemented framework. In order to illustrate how the proposed system works and the kind of natural language generation texts it provides to the users (designers/architects), we introduce a case study of three-dimensional scene synthesis. In particular, the case study illustrates a typical design problem within the selected application domain, that of architectural design of buildings. Such a case incorporates the type of semantic knowledge and the kind of complexity that a reverse declarative design system such RS-MultiCAD can confront.

Table 1. Reflective relations

Object	Relation
Building	longer_than_wide
Garage	wider_than_long

Table 2. Properties

Object	Property	Value
Garage	is_tall	Low
Flat	is_tall	Low
Bathroom	is_wide	Low

The applied scene presents the development of a habitation building. The scene provides a moderate degree of complexity, in a way that it includes objects like "Building", "Garage", "Flat", and "Roof". Also it has three types of the object "Room" that one of "Kitchen", one of "Bathroom", and one of "Bedroom". The number of the different type of relations is eleven, (Tables 1, 2, and 3).

Table 3. Spatial relations

Object	Relation	Object
Building	higher_than	Garage
Roof	adjacent_over, equal_length, equal_width	Flat
Kitchen	adjacent_west, equal_length, equal_width	Bedroom
Bathroom	adjacent_east, equal_length, equal_width	Bedroom
Bedroom	longer_than	Kitchen
Bedroom	longer_than	Bathroom

Additionally the degree of complexity is related in an analogous way with different types of the applied modifications. In the case study the designer-architect implies modifications that altering the geometry of the participating objects. The current example illustrates a site with a building and a garage inside. The building is further decomposed into a flat and a roof. The flat consists of a kitchen, bedroom and bathroom. Tables 1, 2, and 3 present the reflective relations, properties and spatial relations that initially constitute the rule set of the example. The working space of the system prototype presents (Fig. 9), in the middle the selected scenes, on the left-hand side the declarative layer of the stratified representation and on the right-hand side the geometric layer of the selected scene. The relations and properties that belong to the rule set are marked in highlighted red color.

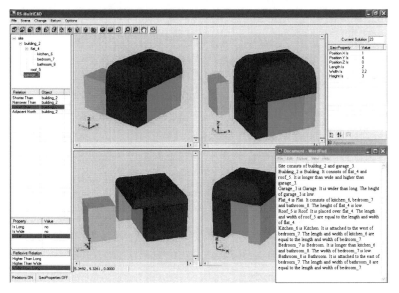

Fig. 9 Initial scenes selection

The designer-architect has the opportunity to select from one to four scenes from the generated solution set. In this case, he/she selects the following four scenes as Fig. 9 presents. The generated text that is based on the initial semantic model has the following form and content:

Site consists of buildng_2 and garage_3.

Building_2 is Building. It consists of flat_4 and roof_5. It is longer than wide and higher than garage_3.

Garage_3 is Garage. It is wider than long. The height of garage_3 is low.

Flat_4 is Flat. It consists of kitchen_6, bedroom_7 and bathroom_8. The height of flat_4 is low.

Roof_5 is Roof. It is placed over flat_4. The length and width of roof_5 are equal to the length and width of flat_4.

Kitchen_6 is Kitchen. It is attached to the west of bedroom_7. The length and width of kitchen_6 are equal to the length and width of bedroom_7.

Bedroom_7 is Bedroom. It is longer than kitchen_6 and bathroom_8. The width of bedroom_7 is low.

Bathroom_8 is Bathroom. It is attached to the east of bedroom_7. The length and width of bathroom_8 are equal to the length and width of bedroom_7.

During the first attempt of the modification, the designer selects the top left scene of the four preselected scenes, and he/she moves the object *"Garage"* to a new position. The system propagates the modification to ancestors and recalculates the valid relations and properties. The modification is valid since there is none relation/property, that belongs to the rule set, which is violated.

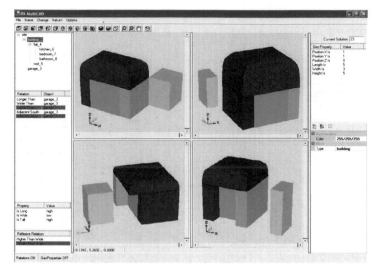

Fig. 10 Top left scene modification

Then, the designer inserts the relation "*building adjacent_west garage*" to the rule set of the specific scene, altering in this way the relative position of the main building with the garage. Fig. 10 illustrates the result of the two modification operations.

The designer selects the top right scene and moves the object "*Building*" to a new position. The system propagates the modification to ancestors, children and recalculates the valid relations and properties. The modification is valid since there is none relation/property, that belongs to the rule set, which is violated. The designer inserts the relation "*Building adjacent_east Garage*" to the rule set of the specific scene. Fig. 11 presents the result of the two operations.

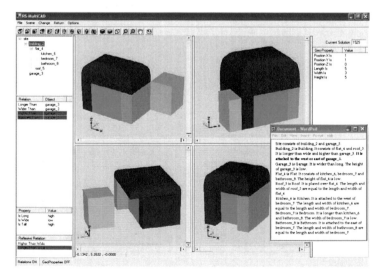

Fig. 11 Top right scene modification

The new generated text embodies the modifications and appears the new relation in bold.

> Site consists of buildng_2 and garage_3.
>
> Building_2 is Building. It consists of flat_4 and roof_5. It is longer than wide and higher than garage_3. **It is attached to the west or east of garage_3.**
>
> Garage_3 is Garage. It is wider than long. The height of garage_3 is low.
>
> Flat_4 is Flat. It consists of kitchen_6, bedroom_7 and bathroom_8. The height of flat_4 is low.
>
> Roof_5 is Roof. It is placed over flat_4. The length and width of roof_5 are equal to the length and width of flat_4.
>
> Kitchen_6 is Kitchen. It is attached to the west of bedroom_7. The length and width of kitchen_6 are equal to the length and width of bedroom_7.
>
> Bedroom_7 is Bedroom. It is longer than kitchen_6 and bathroom_8. The width of bedroom_7 is low.
>
> Bathroom_8 is Bathroom. It is attached to the east of bedroom_7. The length and width of bathroom_8 are equal to the length and width of bedroom_7.

The designer selects the bottom left scene and he/she resizes the object "*Building*" to a new height. The system propagates the modification to ancestors, children and recalculates the valid relations and properties. The modification is valid since there is none relation/property, that belongs to the rule set, which is violated.

Afterward, the designer adds the property "*Building height equals to 5*" to the rule set of the specific scene. Fig. 12 presents the result of the operation.

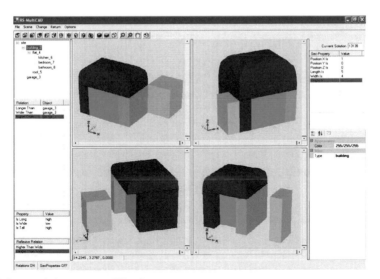

Fig. 12 Bottom left scene modification

In the following step, the designer selects the bottom right scene and he/she increases the property height of the object *"Building"*. The system propagates the modification to ancestors, children and recalculates the valid relations and properties. The modification is valid since there is none relation/property, that belongs to the rule set, which is violated. Next, the designer adds the property *"Building height equals to 6"* to the rule set of the specific scene. In Fig. 13 it is illustrated the result of all the consecutive operations.

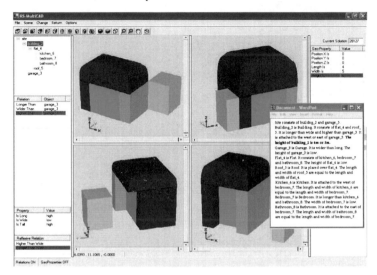

Fig. 13 Bottom right scene modification

The new generated text has the following content:

Site consists of buildng_2 and garage_3.

Building_2 is Building. It consists of flat_4 and roof_5. It is longer than wide and higher than garage_3. It is attached to the west or east of garage_3. **The height of building_2 is 6m or 5m.**

Garage_3 is Garage. It is wider than long. The height of garage_3 is low.

Flat_4 is Flat. It consists of kitchen_6, bedroom_7 and bathroom_8. The height of flat_4 is low.

Roof_5 is Roof. It is placed over flat_4. The length and width of roof_5 are equal to the length and width of flat_4.

Kitchen_6 is Kitchen. It is attached to the west of bedroom_7. The length and width of kitchen_6 are equal to the length and width of bedroom_7.

Bedroom_7 is Bedroom. It is longer than kitchen_6 and bathroom_8. The width of bedroom_7 is low.

Bathroom_8 is Bathroom. It is attached to the east of bedroom_7. The length and width of bathroom_8 are equal to the length and width of bedroom_7.

5 Conclusion

Our motivation is to enable a linkage between two fields, that of declarative modeling and natural language generation, and provide users (in our case designers/architects) with a mutual view of scene synthesis and development based on three-dimensional representation and verbal text. In particular, the basic idea is to link three-dimensional graphics with natural language in order to aid the synthesis and development of complex scenes within fields like virtual reality, computer-aided-design, game design, et cetera.

In the proposed system the reconstruction phase has been introduced in the declarative design cycle in order the design process to be iterative and selected scenes can be modified diversely, leading to solutions closer to requirements. The semantic model is build to support the designer modifications through successive iterations.

The pre-selected scenes are visualized and the semantic model generates an output text that unifies all selected scenes. The generated text facilitates the designer to understand the visualized scenes further even more in cases where the scenes are complex. The resulted system implementation enables proper designer/architect involvement and it eliminates a gap between the original scene description and the resulted updated and modified scenes.

6 Future Work

The horizon of the future work concentrates on the following possibilities. The introduction of more complex scenes, and adding the dimension of temporal alteration, could provide interesting ways to enhance the capabilities of the system. We can adjust the application domain towards domains with demanding three-dimensional graphic design, like game development, engineering, and enchasing the value of the knowledge base by embodying ontology concepts. Besides, the resulted text could be further enhanced by having different prose, in order to be understood by non-experts.

References

[1] Plemenos, D.: A contribution to study and development of scene modeling, generation and display techniques – the MultiFormes project, Professorial dissertation, Nantes, France (November 1991)
[2] Plemenos, D.: Declarative modeling by hierarchical decomposition. In: The Actual State of the MultiFormes project, International Conference Graphic. Con. 1995, St. Petersburg, Russia (July 1995)
[3] Golfinopoulos, V., Bardis, G., Makris, D., Miaoulis, G., Plemenos, D.: Multiple scene understanding for declarative scene modeling. In: 3IA 2007 International Conference on Computer Graphics and Artificial Intelligence, Athens, Greece, pp. 39–49 (2007); ISBN 0-7695-3015-X

[4] Russell, S., Norving, P.: Artificial Intelligence: A modern approach. Prentice Hall, New Jersey (2003)

[5] Miaoulis, G.: Contribution à l'étude des Systèmes d'Information Multimédia et Intelligent dédiés à la Conception Déclarative Assistée par l'Ordinateur Le projet Multi-CAD, Ph.D. Thesis, University of Limoges, France (2002)

[6] Plemenos, D., Tamine, K.: Increasing the efficiency of declarative modeling. Constraint evaluation for the hierarchical decomposition approach. In: International Conference WSCG 1997, Plzen, Czech Republic (1997)

[7] Makris, D.: Aesthetic–Aided Intelligent 3D Scene Synthesis. In: Miaoulis, G., Plemenos, D. (eds.) Intelligent Scene Modelling Information Systems. SCI, vol. 181. Springer, Heidelberg (2009) ISBN 978-3-540-92901-7

[8] Ravani, I., Makris, D., Miaoulis, G., Constantinides, P., Petridis, A., Plemenos, D.: Implementation of architecture-oriented knowledge framework in MultiCAD declarative scene modeling system. In: 1st Balcan Conference in Informatics, Greece (2003)

[9] Dragonas, J.: Modélisation déclarative collaborative. Systèmes collaboratifs pour la modélisation déclarative en synthèse d'image, Ph.D. Thesis, University of Limoges, France (June 2006)

[10] Bardis, G.: Intelligent Personalization in a Scene Modeling Environment. In: Miaoulis, G., Plemenos, D. (eds.) Intelligent Scene Modelling Information Systems. SCI, vol. 181. Springer, Heidelberg (2009) ISBN 978-3-540-92901-7

[11] Ravani, I., Makris, D., Miaoulis, G., Plemenos, D.: Concept-Based declarative description subsystem for CADD. In: 3IA 2004 International Conference, Limoges, France (2004)

[12] Miaoulis, G., Plemenos, D., Skourlas, C.: MultiCAD Database: Toward a unified data and knowledge representation for database scene modeling. In: 3IA 2000 International Conference, Limoges, France (2000)

[13] Golfinopoulos, V.: Understanding Scenes. In: Miaoulis, G., Plemenos, D. (eds.) Understanding Scenes. SCI, vol. 181. Springer, Heidelberg (2009); ISBN 978-3-540-92901-7

[14] Sagerer, G., Niewmann, H.: Semantic networks for understanding scenes. Plenum Press, N. York (1997)

[15] Reiter, E., Dale, R.: Building Natural Language Generation Systems. Cambridge University Press, Cambridge (2000)

[16] Mellish, C., Evans, R.: Implementation architectures for natural language generation. Natural Language Engineering 10(3/4), 261–282 (2004)

[17] Lu, W., Ng, H.T., Lee, W.S.: Natural language generation with tree conditional random fields. In: Conference on Empirical Methods in Natural Language Processing, pp. 400–409 (2009)

[18] Garoufi, K., Koller, A.: Automated planning for situated language generation. In: 48th Annual Meeting of the Association for Computational Linguistics, Uppsala, Sweden, pp. 1573–1582 (2010)

[19] Paris, C., Coloineau, N., Lampert, A., Vander Linden, K.: Discourse planning for information composition and delivery: a reusable platform. Natural Language Engineering 16(1), 61–98 (2009)

[20] Mellish, C., Scott, D., Cahill, L., Paiva, D., Evans, R., Reape, M.: A reference architecture for natural language generation systems. Natural Language Engineering 12(1), 1–34 (2006)

[21] O'Donnell, M., Mellish, C., Oberlander, J., Knott, A.: ILEX: an architecture for a dynamic hypertext generation system. Natural Language Engineering 7(13), 225–250 (2001)

[22] Kosseim, L., Lapalme, G.: Choosing rhetorical structures to plan instructional texts. Computational Intelligence 16(3), 408–445 (2000)

[23] Davey, A.: Discourse Production: A Computer Model of Some Aspects of a Speaker. Edinburgh University Press, Edinburgh (1979)

[24] Goldberg, E., Driedger, N., Kittredge, R.: Using natural-language processing to produce weather forecasts. IEEE Expert 9(2), 45–53 (1994)

[25] Lavoie, B., Rambow, O., Reiter, E.: A fast and portable realizer for text generation. In: Proceedings of the Fifth Conference on Applied Natural Language Processing, pp. 265–268 (1997)

[26] Paris, C., Colineau, N., Lu, S., Vander Linden, K.: Automatically Generating Effective Online Help. International Journal on E-Learning 4(1), 83–103 (2005)

[27] Ren, F., Du, Q.: Study on natutal language generation for spatial information representation. In: 5th International Conference on Fuzzy Systems and Knowledge Discovery, pp. 213–216 (2008)

[28] Karberis, G., Kouroupetroglou, G.: Transforming Spontaneous Telegraphic Language to Well-Formed Greek Sentences for Alternative and Augmentative Communication. In: Vlahavas, I.P., Spyropoulos, C.D. (eds.) SETN 2002. LNCS (LNAI), vol. 2308, pp. 155–166. Springer, Heidelberg (2002)

[29] Kojima, A., Tamura, T., Fukunaga, K.: Natural Language Description of Human Activities from Video Images Based on Concept Hierarchy of Actions. International Journal of Computer Vision 50(2), 171–184 (2002)

[30] Sellinger, D.: Le modélisation géométrique déclarative interactive. Le couplage d'un modeleur déclaratif et d'un modeleur classique, Thèse, Université de Limoges, France (1998)

Objects Description Exploiting User's Sociality

Nikolaos Doulamis[1], John Dragonas[3], Dora Pliota[4], George Miaoulis[2,3], and Dimitri Plemenos[2]

[1] National Technical University of Athens
Herron Polytechniou Str. 15773, Zografou
Athens, Greece
[2] XLIM – UMR 6172 – CNRS,
Department of Mathematics and Computer Sciences
123, avenue Albert Thomas - 87060 LIMOGES CEDEX, France
plemenos@numericable.com, georges.miaoulis@xlim.fr
[3] Technological Education Institute of Athens
Department of Informatics
Ag.Spyridonos St., 122 10 Egaleo, Greece
ndoulam@cs.ntua.gr, idrag@teiath.gr, gmiaoul@teiath.gr
[4] Technological Education Institute of Athens
Department of Business Administration
Ag. Spyridonos St., 122 10 Egaleo, Greece
dpliota@teiath.gr

Abstract. Declarative scene modelling is a very useful modelling technique which allows the user to create scenes by simply describing their wished properties and not the manner to construct them. In declarative modelling, solution filtering is a very important aspect due to the imprecise description of both the scene and the objects, as well as due to the subjective of humans regarding the content of a design is concerned. Currently, solution filtering is performed by the application of machine learning strategies or clustering methods in a collaborative or not framework. However, the main difficulty of these algorithms is that solution filtering is based on the usage of low-level attributes that describe either the scene or the object. This chapter addresses this difficulty by proposing a novel social oriented framework for solution reduction in a declarative modelling approach. In this case, we introduce semantic information in the organization of the users that participates in the filtering of the solutions. Algorithms derived from graph theory are presented with the aim to detect the most influent user with a social network (intra-social influence) or within different social groups (inter-social influence). Experimental results indicate the outperformance of the proposed social networking declarative modelling with respect to other methods.

Keywords: Social networking, declarative modelling, computer graphics, architect design.

1 Introduction

Declarative modeling in computer graphics is a very powerful technique. It allows users to describe the scene in an intuitive manner, by only giving some expected properties of the scene and letting the modeller find solutions, if any, verifying these properties. As the user may describe a scene in an intuitive manner, using common expressions, the described properties are nearly always imprecise.

Traditionally, declarative modeling is implemented under a centralized framework that allows for a single user to describe, retrieve, and evaluate a set of 3D objects, composing a scene. However, due to the recent advances in software and communications technologies, collaborative systems become much easier, efficient and effective, increasing the mutual cooperation between individuals or groups of common objectives [1]. This stimulates the transition from the centralized to the distributed and collaborative declarative modelling framework. Collaborative modelling is actually a recursive process that allows two or more designers or organizations to work together to realize shared goals. It has been experimentally noticed that any task could be achieved much faster and with better results in a collaborative or on a network environment, when the members of a group execute the work in parallel [2].

Although many works have been reported in the literature for collaborative product design [3][4][5], the term "collaboration" is still interpreted differently with respect to a) the technological means used to perform the collaboration and b) the degree of co-operation among the users. For instance, there are collaborative architectures that are based on the monolithic usage of the e-mail service for communication and data sharing among the users, or other that exploit Web based technologies [2].

As the Internet and the Web is revolutionized to a social networking platform and mass communication, social networking tools are emerged that allows people to work together, collaborate with each other and creating value under a transparent and truly co-operative way. Today, many sites, such as Friendster, Facebook, Orkut, LinkedIn, Bebo, and My-Space, as well as content-sharing applications offer social networking functionality [6].

Social networking is a concept that has been around much longer than the Internet or even mass communication. It is within people's behavior to work together, to live with others, and to create groups of common interest the value of which may be much greater than the sum of its parts. Nowadays, Internet is moving, not only towards the media dimension, but also towards the social aspect. This means in other words that the traditional Media Cyberspace is transforming from the traditional top-down paradigm of few large media corporations creating content for the consumers to access, the production model to that of individualized content now being created by everybody [7], [8]. While we previously thought of the Internet as an information repository, the advent of social networks is turning it into a tool for connecting people. The mass adoption of social-networking websites of all shapes and sizes points to a larger movement, an evolution in human social interaction.

Introducing social networking aspects into the collaborative declarative systems is an open research agenda in the area of computer graphics and image processing. This is the main research objective of this chapter; to propose a new social based declarative modeling architecture.

In general, the declarative scene modeling is based on a declarative conception cycle, which consists of three sequential function phases [1]:

- *The scene description cycle* that describes the properties relating to the description of a scene.
- *The Solutions generation cycle* which is responsible for the retrieving relevant solutions in accordance with the requested scene descriptions
- *The scene understanding cycle* that allows users to evaluate the precision of the generated (retrieved) solutions.

In this chapter, we focus on *the solutions generation cycle*, that is, on the way of retrieving 3D objects of interest that satisfies user's information needs.

1.1 Previous Works

Currently, 3D objects are retrieved by matching the objects' attributes along with user's preferences (query requests), defined in the scene description cycle. However, there are many difficulties that prevent the system to retrieve and display the most appropriate solutions to the designers. One difficulty is the lack of precision in describing a scene and/or a 3D object. For example, the expression "put scene A on the left of scene B" imposes several possibilities on the way that the scene A is put on the left of the scene B. Another difficulty stems from the fact that the designer does not know the exact property of the scene. For example the statement "Object A should be near to Object B" may have many interpretations by the modeller. Finally, there is *subjectivity* as far as the content of a design is concerned; different users (designers) express their views in different ways.

To address the aforementioned difficulties, a machine learning methodology is proposed in [9]. In this work, a neural network structure is introduced to dynamically model user preferences in order to yield to a reduction of the possible solution spaces. In this way, we can filter the generated solutions to fit user's preferences. However, since user's preferences are dynamic and social context are dynamically evolving, adaptable methodologies [11] are required for user profile estimation and social context understanding [10].

However, the main drawback of [9] is that solution filtering is performed without taking into account any co-operation among the users. This difficulty is addressed in [12] where a collaborative mechanism is proposed for the solution filtering. The method for preferences integration proposed in [12] takes into account the independent dimensions of group policy and member importance. It allows the adjustment of the consensus mechanism, thus offering alternative configurations to the design group coordinator.

More sophisticated approaches are presented in [13]. In this work, we have exploited advanced clustering methodologies for solution filtering under a collaborative framework. In particular, we have proposed a spectral clustering technique [14] which personalizes the retrieval of the relevant solutions taking into account dependencies among them and users' preferences. Spectral clustering is a very powerful method for graph partitioning and can be used in many other application scenarios like workflow management in service engineering [15]. Other methods include the usage of artificial intelligence, like on-line learning strategies [16] and neural networks methods [17].

1.2 Contribution

All the aforementioned approaches filter the generated solutions by exploiting either single user's preferences or collaboratively combining the preferences of multiple users. However, solution filtering is accomplished on low level features regarding the description of the 3D objects (shape, texture, location, etc). Although the capabilities of the previously proposed automatic clustering methodologies, like, for example, the use of spectral clustering, or the introduction of non-linear schemes for on-line learning of user's preferences, the exploitation of low-level descriptors set insurmountable obstacles in the actual scene understanding as far as users' is concerned. For instance affective/ decorative description of a scene cannot be modeled by the use of low-level descriptors even the best clustering algorithms are implemented.

All we are living within social communities. We are thinking and behaving with respect to cultural and social constraints, which actually define our preferences. All these constitute a high level description of a scene, which is far away from the low-level description techniques adopted so far. This means that solution filtering should be accomplished in a *social dimension,* instead of a *preference based dimension* even if the latter is implemented under a collaborative framework.

This is the main contribution of this chapter; to move from a preference oriented solution filtering to a *social dimension declarative modelling.* Social aspects in declarative modelling mean that when we are collaborated with others we trust more the opinion of the social group, to which we share common ideas, instead of other users.

Under this context, we propose methodologies able to estimate the influence node (user) of a social network by exploiting issue derived from the graph theory. We estimate the prominent user regarding the eigenvalue Centrality, which is a measure for estimating node connectivity. Then, we apply an algorithm for reducing solutions of a design according to the properties of the social network of a user as indicated by the prominent user.

This chapter is organized as follows: Section 2 presents the concept of the proposed social declarative modelling, while Section 3 discusses issues regarding the solution filtering according to the properties of the social network. In this section, we present methods for estimating the most influent (prominent) node (user) within a social network. In addition, we present the algorithm for solution reduction according to the properties of the social group. Section 4 presents simulation results regarding the proposed methods along with other approaches. Finally Section 5 concludes this chapter.

2 Social Declarative Modelling

Humans are social beings! According to Aristotle, "Man is by nature a social animal; an individual who is unsocial naturally and not accidentally is either beneath our notice or more than human". Thus, it is not surprising to take in order to social aspects into 3D content services adaptability. Now computers and the Web are the natural media for a wide spectrum of new, inherently social activities. Thus, in this chapter, we investigate socially aware mechanisms in the declarative framework environment [18].

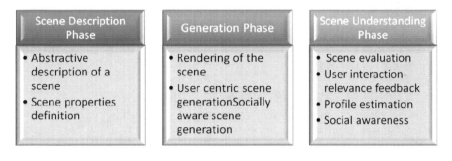

Fig. 1 The main components of a declarative modelling architecture

2.1 Declarative Modelling Principles

As we have stated in section 1, the design can be improved if collaboration of different designers on the same product is accomplished. This is mainly due to the fact that contemporary scene design problems are inherently complex and require multiple designers to work collaboratively. The need for collaboration appears when individuals fail to carry out a task on their own because of their limited abilities (lack of knowledge or strength), or when collaboration can help them to carry out this task faster and more efficiently [19]. It has been experimentally noticed that any task could be achieved much faster and with better results in a collaborative or on a network environment, when the members of a group execute the work in parallel [19], [20]. The declarative modelling includes three main phases; *the scene description phase, the generation phase* and the *scene understanding phase* (see Section 1).

Figure 1 presents the main structure of the declarative modelling principles. In declarative modelling, scene is described in an abstractive way. This is performed during the scene description phase. In this phase, we also define the properties of the scene. Abstractive description has the advantage that the scene is defined with respect to the semantics (meanings) while implementation details are hidden. Abstraction tries to reduce and factor out details so that the designer can focus on a few concepts at a time. A system can have several abstraction layers whereby different meanings and amounts of detail are exposed to the designer [21]. The main disadvantage of an abstractive description is that humans are unique and they interpret a concept using with high-level subjective criteria which may quietly vary from person to person or even the same person under different circumstances. This yields in an erroneous scene generation from user's point of view. In other words, the smarter scene generator is not able to produce scenes that satisfy the current user's information needs by taking as inputs only an abstract description of a scene, since the same description have quite different meanings from different people (humans' subjectivity).

To address these difficulties, a scene understanding phase has been included in the declarative scene modelling. Scene understanding allows user's interaction with the system and evaluation of the generated scenes. This way, we are able to estimate the basic user's profile space, which are then used to adapt the generated scenes according to user's preferences.

2.2 Social Networking

Social identity theory provides an appropriate framework for analyzing social aspects of collaborative technology design, in which design team members have different "design identities," that is, sets of beliefs, attitudes, and values about design. Observations of novice and professional design teams support the notion that collaboration is influenced by designers' adherence to, and reaction against, idealized technology-centred and socially-centred categorizations of themselves and users [22].

Social networking is a concept that has been around much longer than the Internet or even mass communication. It is within people's behaviour to work together, to live with others, and to create groups of common interest the value of which may be much greater than the sum of its parts. Nowadays, Internet is moving, not only towards the media dimension, but also towards the social aspect. This means in other words that the traditional Media Cyberspace is transforming from the traditional top-down paradigm of few large media corporations creating content for the consumers to access, the production model to that of individualized content now being created by everybody. While we previously thought of the Internet as an information repository, the advent of social networks is turning it into a tool for connecting people. The mass adoption of social-networking websites of all shapes and sizes points to a larger movement, an evolution in human social interaction.

Within the past decade, Internet traffic has shifted dramatically from HTML text pages to multimedia file sharing as illustrated by the emergence of large-scale multimedia social network communities such as Napster, flickr, and YouTube [23]. For example, a study showed that in a campus network, peer-to-peer file sharing can consume 43% of the overall bandwidth, which is about three times of all WWW traffic [24]. This consumption poses new challenges to the efficient, scalable, and robust sharing of multimedia over large and heterogeneous networks. It also significantly affects the copyright industries and raises critical issues of protecting intellectual property rights of multimedia [25].

In the popular press, the term social network usually refers to online communities such as Facebook, MySpace, or Friendster. In the broader scientific literature, it is often associated with research that took off after the appearance of an article on small worlds in Nature [26] and one on preferential attachment in Science [27].

The term was issued, however, in a 1950s article [28] to signify a perspective different from those focusing on individuals, social groups, or social categories.

Today, social network analysis has matured into a paradigm, combining structural theories of social interaction with corresponding methods for its analysis [29]. A social network may represent any form of relation (affection, dependency, power, support, etc.) between any type of social actors (people, organizations, nations, etc.), potentially through mediating entities (joint participation, shared beliefs, etc.). In an analysis it can be a dependent ("Why birds of a feather flock together?") or an explanatory variable ("Are people with influential friends more powerful?"). A somewhat distinct line of research is concerned with personal networks, where each network is considered an attribute of a focal actor defining its boundaries. Of course, social networks are subject to measurement [18].

On the other hand, face-to-face communication conveys social context as well as words. It is this social signalling that allows new information to be smoothly integrated into a shared, group-wide understanding. Social signalling includes signals of interest, determination, friendliness, boredom, and other "attitudes" toward a social situation [30]. Social signals and tools allow us to predict human behaviour and sometimes exceed even expert human capabilities. These tools potentially permit computer and communications systems to support social and organizational roles instead of viewing the individual as an isolated entity. Example applications include automatically patching people into socially important conversations, instigating conversations among people in order to build a more solid social network, and reinforcing family ties.

In this chapter, we include the concept of social networking into the declarative modelling scenarios. In other words, we improve declarative modelling taking into account social aspects, that is, the knowledge of friends, family and other group members that share similar beliefs, attitudes, and views with us on a live event performance. We can thus "filter out" un-necessary content, we can adapt the stream information to our personal preferences we can select best results relying on the knowledge derived from our social group, or even better trusting more the suggestions of our fellows in the group.

Requirements: To integrate social aspects into a declarative modelling architecture , we need (a) monitoring of the social activities of the users so as to extract their social profile, (b) conclusions regarding the services parameters as being suggested by the social group, [i.e., quality for the generated scenes content delivery delay, type of interesting content, speed in media transmission (off-line, just-in-time, real-time), etc], (c) automatic discovery (localization) who of the social members are the closers users (in the sense that they perform as similar as possible) to an individual (exploring, for instance, social graphs) and (d) and automatic search for enhanced virtual/real users' generated content that can enriched the live captured one.

We address all these issues in this chapter. For this reason, in the following sector, we give an overview of the developed scene and how one should modify declarative modelling as that is has been tailored to user's social needs.

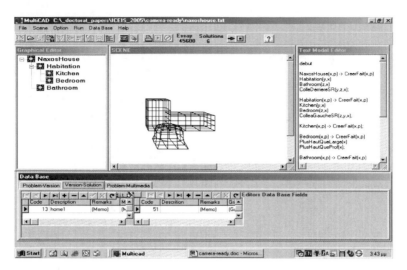

Fig. 2 A Typical MultiCAD Session

2.3 The MultiCAD Architectural Framework

MultiCAD is an intelligent multimedia information system that is based on declarative modelling, developed by the Laboratory Méthodes et Structures Informatiques of the University of Limoges along with the Team of Information Systems & Applications of the Department of Informatics of the Technological Educational Institute of Athens. It was used for evaluating the proposed social aware declarative modelling architecture. Figure 2 presents a typical session of the MultiCAD framework.

2.4 Overview of the Proposed Social Declarative Modelling System

Figure 3 proposes the architectural structure of the existing methodologies [12], [13] for solutions filtering in a declarative collaborative design framework. Despite the algorithm used to filter out the solutions, initially the profile of the users is extracted under a collaborative way and then the generated solutions are tailored to user's preferences constraints before visualizing the results and depicting them to the users. In addition, each constrained solution to user's preferences is also reused to estimate better profile or even to capture the temporal dynamics of the profile since users' needs may change from time to time. This way the retrieved solutions are framed with respect to user's preferences and his/her profile assisting him/her in proper selection 3D objects during a design.

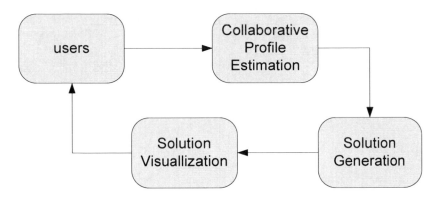

Fig. 3 The main modules of the existing methodologies for solution filtering with respect to user's preferences [9], [12], [13]

As mentioned above, profileration is performed on the basis of low-level descriptors. This implies that more complicated concepts involved in the selection process cannot be taken into account regardless of the efficiency of the algorithm used to estimate the profiles and/or representations of the 3D objects. For instance, the artistic, decorative position of an object cannot be properly modelled exploiting the low-level attributes. Furthermore, a designer cannot select 3D objects with respect to their affective effect in the scene or even based on designer's cultural interpretation about the content.

This is addressed, in this chapter, since we are moving from the conventional collaborative declarative modelling towards a social-based description of a scene that could fulfil and skip all the above mentioned obstacles. What we introduce in this chapter is the inclusion of the social profile estimation in the architectural loop. Figure 4 shows how the components of Figure 3 are modified to incorporate

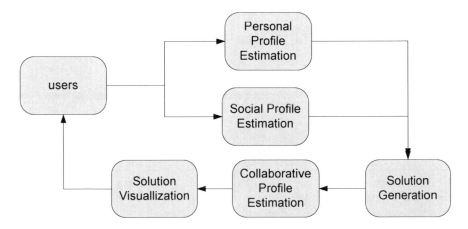
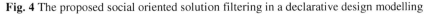

Fig. 4 The proposed social oriented solution filtering in a declarative design modelling

social aspects in the design phase during the solution generation cycle. In particular, the users are said to belong into social communities and via them the social profile of the users is estimated.

Then, a set of new solutions is generated properly constrained on social group preferences.

3 Social Filtering

Let us assume that a user belongs to a social group. This group is constructed according to a set of semantics that describe user's profile. It is clear that a social group has specific interest regarding scene description and object selection. Therefore, individuals of a social group follows the interest of the "influence" social node (user), without, however, loss the specific individualized properties of the users [18].

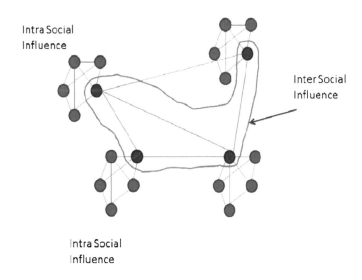

Fig. 5 The Intra and the Inter Social influence

3.1 Social Graphs

A social network is defined as a set of people or group of people with some pattern of interrelations between them. A social network can be mathematically modelled using concepts from graph theory. A graph is a collection of vertices (or nodes) and edges between them. The vertices are abstract nodes and edges represent some sort of relationship between them. In the case of a social network the vertices are people and the edges represent a kind of social or person-to-person relationship.

In particular, we represent a social network as a graph $G = \{V_i, E_i\}$, where V_i represents the nodes (vertexes) of the social networks and E_i the relationship (graph edge) between the two nodes. In a social network, edge E_i corresponds, for example to the "friendship" or "similarity" between two graph nodes that this edge connects.

Every graph is represented by its respective adjacent matrix. In the following, we denote this matrix as A_G to indicate its relationship with the graph $G = \{V_i, E_i\}$. The A_G is an NxN square matrix, where variable N represents the number of nodes (vertexes) of the social network. The elements $a_{i,j}$ of the matrix $A_G = [a_{i,j}]$ correspond to the relationship between the *i-th* and *j-th* node of the graph. In this way, we can apply concepts and tools for linear algebra for analyzing a social network and estimate the most influent user (or graph node).

Two types of social influence are proposed in the proposed chapter. The first concerns the *Intra-Social influence*, while the second the *Inter-Social influence*. Intra-social influence finds the most relevant influence node (user) with the same social class. On the contrary, inter social influence finds the most prominent user for classes different than the one the user belongs to. This concept is illustrated in the Figure 5.

3.2 Intra - Social Influence

Using the mathematical formulation of a graph, the question that arises is how we can estimate the social influence or important node within the graph.

3.2.1 Social Influence as Connectivity

One way to measure influence is connectivity. People who have lots of friends tend to have more influence (indeed, it's possible they have more friends precisely because they are influential). Using the adjacent matrix A_G, we can easily measure the connectivity of a node (user) as the degree of a node in the graph; the number of other nodes to which it is connected.

In particular, let as denote as $I_c(p)$ the influence metric with respect to node connectivity for a user p. Then, we can define this measure as

$$I_c(p) = \# \text{ of friends for user } p \qquad (1)$$

Then, modelling the social network as the adjacent matrix, we can estimate metric $I_c(p)$ of (1) as the row-sum of the matrix A_G for the row p.

$$I_c(p) = \sum_{i=1}^{N} a_{i,p} \qquad (2)$$

The main drawback of using equation (2) is that connectivity is not always a good metric for estimate the influence of a user to the social network. This is depicted in the Figure 6. It is clear that the node (user) p_1 has a higher measure of influence than the node p_2 because they are directly connected to eight people. However, the second node (e.g., the p_2) has the potential to be connected up to 9 users. Therefore, node p_2 seams to be more representative than the node p_1.

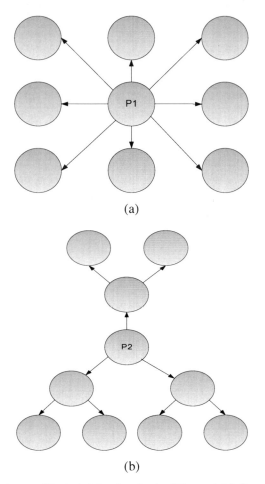

(a)

(b)

Fig. 6 Two different cases illustrated the drawback of the social influence as connectivity. The node p1 has a higher measure of connectivity than the node p2. However, node p2 seams to be more representative than the node p1

3.2.2 Eigenvalue Centralirity

One way to handle the aforementioned drawback is to use the so-called *eigenvalue (or eigenvector) centrality*. The idea is to propagate the influence of a node proportionally to the degree of relationship that this node connects with other nodes.

Let us denote as $I_e(p)$ the so-called node influence with respect to the eigenvalue cardinality. Then, using the values of the adjacent matrix A_G, we can write relate variable $I_e(p_i)$ with the influence degrees of the other associated nodes $I_e(p_i)$ as follows

$$I_e(p_i) = \frac{1}{\lambda} \sum_{j=1}^{N} a_{i,j} I_e(p_j) \qquad (3)$$

Equation (3) means that the degree of influence for a node p_i is associated with the degree of influence for all nodes p_j that this node is connected with.

In the following, we re-write equation (4) in a matrix form by taking into consideration all the nodes p_i of the social network. In particular, let us denote as **x** a vector that contains the degree of influence for all the users in the social network. Then, we have that

$$\mathbf{x} = [I_e(p_1)\ I_e(p2)\ I_e(p_3)....]^T \qquad (4)$$

Using the aforementioned equation, we can re-write equation (5) as follows by taking into consideration all the nodes of the social network

$$\mathbf{x} = \frac{1}{\lambda} \mathbf{A}_G \cdot \mathbf{x} \qquad (5)$$

where **A** refers to the adjacent matrix of the social network.

It is clear that equation (6) can be written as an eigenvalue problem,

$$\lambda \mathbf{x} = \mathbf{A}_G \cdot \mathbf{x} \qquad (6)$$

Therefore, calculating someone's influence according to $I_e(p)$ is equivalent to calculating what is known as the "principal component" or "principal eigenvector."

3.2.3 Detecting the Influence Node

One simple way for Detecting the influence node of a social graph is to estimate the dominant eigenvector of the adjacent matrix A_G that represents the social network. Then, the most influent node is the one that corresponds to the largest value of the dominant eigenvector.

A very efficient way for calculating the principal eigenvector is the so-called power method. The idea is that if you take successive powers of a matrix A_G, normalize it, and take successive powers then the result approximates the principal eigenvector.

The algorithm starts with an equal value for all the nodes as far as the eigenvalue influence is concerned. This means that

$$\mathbf{x}^{(0)} = [\frac{1}{N}\ \frac{1}{N}\ \frac{1}{N}....]^T \qquad (7)$$

Then, the values are iteratively calculating using the power law method formula

$$\mathbf{x}^{(k)} = \mathbf{A}_G \mathbf{x}^{(k-1)} \tag{8}$$

It can be easily proven that equation (8) converge to the principal eigenvector of the matrix A_G. Therefore, equation (8) is a good estimate for the eigenvalue influence of the nodes of the social graph, that is of the vector x.

3.3 Inter Social Influence

In the following section, we describe the intra social influence of the nodes of a graph. In particular, the node that has the largest eigenvalue influence is selected as the most representative node among the social network. However, such approach does not take into consideration the inter-relationships among different social networks. These inter-relationships may significantly improve the performance of the solutions returned to the end users.

The inter-social influence is approximated using the most influent nodes of a social graph. In particular, let us assume that we have K social networks, $G_k = \{V_i^{(k)}, E_i^{(k)}\}$. Each social network is represented as a graph. Using the power law method, we can estimate the most representative (influent node) of this social graph. Let us denote as $n_l^{(k)}$ the influent node for the k-th social network. Then, an inter-social graph is constructed by taking into account only the influent (representative) nodes for each social network. We denote this inter-graph as $G_{\text{inter}} = \{V_i^{\text{inter}}, E_i^{\text{inter}}\}$. Variable V_i^{inter} represents the influence nodes of each social network, while E_i^{inter} the relationship between two influent nodes.

3.3.1 Detection of Inter-influence Node

Using the previous formulation, the inter-social graph is represented using the adjacent matrix A_G^{inter}. Then, we can use the power law methodology, as described in Section 3.2, for estimating the degree of eigenvalue influence for each representative node in the inter-social network.

It is clear that in the inter social solution filtering, our goal is not to estimate the principal influence node of the graph. Instead, our purpose is to estimate the degree of influence in order to re-adjust the solutions according to these degrees.

3.4 Social Solution Filtering

Let us assume that each object is associated with a number of attributes that represent the properties of the object. These attributes are, for example, shape, texture and geometric characteristics of the object. We denote these attributes as $\mathbf{a} = [a_1, a_2, ...]^T$. For each of these attributes, we associate a degree of important (weight). The attribute weights are included in the weight vector, say $\mathbf{w} = [w_1, w_2, ...]^T$. It is clear that solution filtering is accomplished by re-weighted the degree of importance for each of the attributes of the objects. However, in this chapter, the weights are estimated according to the influence node (user) of the social network that this user belongs to.

In particular, let us suppose that each user within the social network is characterized by degree of importance vector, say $\mathbf{w}(p)$. Using the values of the vector $\mathbf{w}(p)$, we can derive the adjacent matrix A_G of the social network. In other worlds, we can estimate a degree of relationship among two users (nodes) of the network. It is clear, however, that this degree is sometimes manually set by the preferences of the users. It is also cleat that this is happen using high level semantic information, which is far away from the low-level attributes that represent the properties of the scene objects. Therefore the relationship between two users (nodes) can be described as follows

$$a_{i,j} = d(\mathbf{w}(p_i), \mathbf{w}(p_j)) + \gamma \cdot S(p_i, p_j) \tag{9}$$

where we define as $d(\mathbf{w}(p_i), \mathbf{w}(p_j))$ the distance between the users i and j, as estimated by the inner product of their respective weight vectors and $S(p_i, p_j)$ the respective social relationship between the i and j, as defined by the users. In equation (9), variable γ is a scalar factor that updates the degree of importance between the low-level attributes and the semantic information provided by the term $S(p_i, p_j)$.

Table 1 The percentage of user's satisfaction for different number of retrieved solutions under different filtering methodologies

Algorithms	Number of Retrieved Solutions		
	20	30	40
Independent user profile modelling	48%	46%	43%
User's preference consensus	52%	49%	46%
Centre based clustering	54%	52%	49%
Spectral clustering	55%	53%	50%
Social Declarative modelling (Intra-Social approach)	63%	62%	58%
Social Declarative modelling (Intra-Social approach)	65%	64%	59%

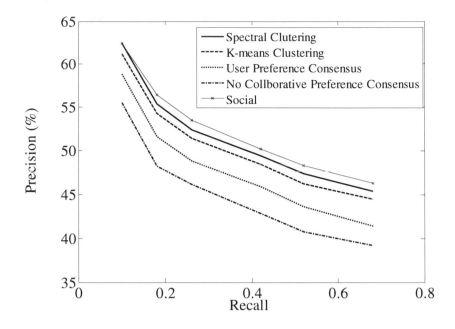

Fig. 7 The precision-recall curve for different solution filtering methods

Initially, the solutions are filtered according to the values of the weight vector of the prominent (influent) node in the social graph. Then, the weight vector of each user in the social graph is updated, resulting in a respective updating of the extracted influent node. This is achieved by introducing a kind of interaction between the proposed solution filtering architecture and the end-user.

4 Simulations

In this section, we present simulation results as far as the proposed solution filtering technique that exploits users' interaction. Table 1 presents the percentage of users' satisfaction regarding the proposed algorithm. In this Table, we have evaluated the results of existing methodologies in the area of user's profile in the declarative modelling. In particular, we present the results in case of independent user profile modelling, that is without the exploitation of preferences from other users in a collaborative design framework. In addition, we present the results of the preference consensus approach, where solution are filtered using the cooperation between different users. Finally, we examine two clustering methods for grouping the users according to the low-level attributes; the centre based clustering which is based on the application of the k-means algorithm and the spectral clustering method which groups users according to the bi-directional relationships. The results have been tested from a architecture design test sequence.

In this table, we have presented the results of the proposed social based clustering. We have illustrated the results regarding either the intra-social or the inter-social algorithms. Although the inter social framework improves, on average, the results, there are cases where the inter-social filtering may reduce the filtering efficiency. This is mainly in cases where the different social networks are quite dissimilar with each other. However, the introduction of inter-social filtering expands the retrieved solutions to other paths outside of the social network. This is a very important aspect, especially in case of architecture design.

It is clear that in all cases our method outperforms the existing methodologies, since it incorporates the semantic information regarding user's grouping.

Another metric of evaluation of the proposed filtering approach is the precision-recall curve. Precision is the fraction of retrieved documents that are relevant to the search. On the other hand, Recall is the fraction of the documents that are relevant to the query that are successfully retrieved. For an "ideal" system, both Precision and Recall should be high (see Figure 7). However, in a real retrieval system, as the number M of images returned to the user increases, precision decreases, while recall increases. Because of this, instead of using a single value of Precision or Recall, the curve Precision-Recall is usually adopted to characterize the performance of a retrieval system.

5 Conclusions

In this chapter, we have presented the concept of the social declarative modelling for an efficient scene design. Collaborative declarative modelling is a very importance research aspect in computer graphics, since it allow the efficient construction of complex architectural designs. However, although many works have been proposed in the literature for collaborative product design, the term "collaboration" is still interpreted differently with respect to the technological means used for the collaboration and the degree of co-operation between the users. Social networking is a new framework that allows people to work together and collaborate with each other under a truly co-operative framework. In this chapter, we have proposed a social oriented collaborative declarative modelling framework using methods from social graphs.

In the declarative modelling, solution filtering is one of the most important aspects due to the vague description both of a scene and of the object description. Therefore, methods for filtering the solutions according to user's preferences constitute one of the most important aspects in the declarative modelling research. However, today, solution filtering is accomplished by the exploitation of the low-level attributes used for representing the objects in a scene. Then, filtering is performed through the application of advanced clustering methods, such as centre based grouping tools and/or spectral clustering.

However, despite the capabilities of these methodologies, the exploitation of low-level features for solution filtering set insurmountable obstacles in the efficiency of these methods. Humans perceive objects and a scene using high level semantics, which are not able to be modelled based on the low-level extracted attributes. For instance, it is not possible to measure the decorative or the artistic degree of an object with respect to a scene.

These difficulties are addressed in this chapter, by proposing a social framework for the collaborative declarative modelling. The social framework allows users of common interest to work and collaborate together. In this case, only users that are close in a semantic dimension are able to participate in solution filtering, instead of the previous methods where the users' grouping is performed under a low-level descriptor framework.

In this chapter, we apply graph theory methods for detecting the influence user within a social network. This influent user is, then, used for solution filtering. Intra and Inter social filtering algorithms are presented. Intra social filtering allows solution reduction by taking the results only from one social network. On the contrary, inter social filtering allows for solution reduction by incorporating aspects from other "similar" social networks.

References

[1] Dragonas, J., Makris, D., Lazaridis, A., Miaoulis, G., Plemenos, D.: Implementation of Collaborative Environment in MultiCAD Declarative Modelling System. In: International Conference on Computer Graphics and Artificial Intelligence (3IA 2005), Limoges, France, May 11-12 (2005)

[2] Dragonas, J., Doulamis, N.: Web-based Collaborative System for Scene Modelling. In: Intelligent Scene Modelling Information Systems, pp. 121–151. Springer, Berlin (2009)

[3] Rodriguez, K., Al-Ashaab, A.: A Review of Internet based Collaborative Product Development Systems. In: Proceedings of the International Conference on Concurrent Engineering: Research and Applications, Cranfield, UK (2002)

[4] Shen, W.: Web-based Infrastructure for Collaborative Product Design: An Overview. In: 6th International Conference on Computer Supported Cooperative Work in Design, Hong Kong, pp. 239–244 (2000)

[5] Kvan, T.: Collaborative design: What is it? Automation in Construction 9(4), 409–415 (2000)

[6] Breslin, J., Decker, S.: The Future of Social Networks on the Internet- the need for semantics. IEEE Internet Computing, 86–90 (November-December 2007)

[7] Ko, M.N., Cheek, G.P., Shehab, M., Sandhu, R.: Social Networks Connect Services. IEEE Computer Magazine, 37–43 (August 2010)

[8] Douglis, F.: It's All About the (Social) Network. IEEE Internet Computing, 4–6 (January-February 2010)

[9] Plemenos, D., Miaoulis, G., Vassilas, N.: Machine Learning for a General Purpose Declarative Scene Modeller. In: International Conference on Computer Graphics and Vision (GraphiCon), Nizhny Novgorod (Russia), September 15-21 (2002)

[10] Doulamis, A.: Dynamic Tracking Re-Adjustment: A Method for Automatic Tracking Recovery in Complex Visual Environments. Multimedia Tools and Applications 50(1), 49–73 (2010)

[11] Doulamis, A.: Adaptable Neural Networks for Objects' Tracking Re-initialization. In: Alippi, C., Polycarpou, M., Panayiotou, C., Ellinas, G. (eds.) ICANN 2009. LNCS, vol. 5769, pp. 715–724. Springer, Heidelberg (2009), doi:10.1007/978-3-642-04277-5_72

[12] Bardis, G., Doulamis, N., Dragonas, J., Miaoulis, G., Plemenos, D.: A Parametric Mechanism for Preference Consensus in a Collaborative Declarative Design Environment. In: International Conference on Computer Graphics and Artificial Intelligent, Athens, Greece (2007)

[13] Doulamis, N., Dragonas, J., Doulamis, A., Miaoulis, G., Plemenos, D.: Machine learning and pattern analysis methods for profiling in a declarative collaorative framework. In: Plemenos, D., Miaoulis, G. (eds.) Intelligent Computer Graphics 2009. Studies in Computational Intelligence, vol. 240, pp. 189–206. Springer, Heidelberg (2009)

[14] Ng, Y., Jordan, M.I., Weiss, Y.: On spectral clustering: analysis and an algorithm. Neural Information Processing Systems 14 (2002)

[15] Delias, P., Doulamis, A., Matsatsinis, N.: A Joint Optimization Algorithm For Dispatching Tasks In Agent-Based Workflow Management Systems. In: International Conference on Enterprise Information Systems (ICEIS), Barcelona, Spain (June 2008)

[16] Rui, Y., Huang, T.S.: Optimizing Learning in Image Retrieval. In: Proc. of IEEE Int. Conf. on Computer Vision and Pattern Recognition (June 2000)

[17] Haykin, S.: Neural Networks: A Comprehensive Foundation, 2nd edn. Prentice Hall Press, Englewood Cliffs (1998)

[18] Brandes, U.: Social network analysis and visualization. IEEE Signal Processing Magazine 25(6), 147–151 (2008)

[19] Dragonas, J., Doulamis, N.: Web-based collaborative system for scene modelling. In: Miaoulis, G., Plemenos, D. (eds.) Intel. Scene. Mod. Information Systems. SCI, vol. 181, pp. 121–151. Springer, Heidelberg (2009)

[20] Doulamis, N., Bardis, G., Dragonas, J., Miaoulis, G.: Optimal recursive designers' profile estimation in collaborative declarative environment. In: International Conference on Tools with Artificial Intelligence, ICTAI 2, art. no. 4410416, pp. 424–427 (2007)

[21] Harold, A., Jay Sussman, G., Sussman, J.: Structure and Interpretation of Computer Programs, 2nd edn. MIT Press, Cambridge (1996)

[22] Kilker, J.: Conflict on collaborative design teams: understanding the role of social identities. IEEE Technology and Society Magazine 18(3), 12–21 (1999)

[23] Gummadi, G.P., Dunn, R.J., Saroiu, S., Gribble, S.D., Levy, H.M., Zahorjan, J.: Measurement, modeling and analysis of a peer-to-peer file-sharing workload. In: Proc. 19th ACM Symp. Operating Systems Principles (SOSP-19), pp. 314–329 (October 2003)

[24] Liang, J., Kumar, R., Xi, Y., Ross, K.W.: Pollution in P2P file sharing systems. IEEE InfoCom 2, 1174–1185 (2005)

[25] Vicky Zhao, H., Sabrina Lin, W., Ray Liu, K.J.: Behavior Modeling and Forensics for Multimedia Social Networks. IEEE Signal Processing Magazine, 118–139 (January 2009)

[26] Watts, D.J., Strogatz, S.H.: Collective dynamics of small-world networks. Nature 393(6684), 440–442 (1998)

[27] Barabási, A.L., Albert, R.: Emergence of scaling in random networks. Science 286(5439), 509–512 (1999)

[28] Barnes, J.A.: Class and committees in a Norwegian Island parish. Human Relations 7(1), 39–58 (1954)

[29] Freeman, L.C. (ed.): Social Network Analysis I–IV. Sage, Newbury Park (2008)

[30] Pentland, A.S.: Social Signal Processing, pp. 108–111 (July 2007)

Knowledge-Based Analysis and Synthesis of Virtual 3D Campus Infrastructure Models

Wolfgang Oertel, Hermin Kantardshieffa, and Maria Schneider

Dresden University of Applied Sciences, Friedrich-List-Platz 1, 01069 Dresden, Germany
{Oertel,Kantards,Mschneid}@informatik.htw-dresden.de
www.informatik.htw-dresden.de

Abstract. The focus of the paper lies on the use of traditional artificial intelligence technologies for a practical application - the support of the analysis and synthesis of three-dimensional models of campus infrastructures. The centre stage is taken by a combination of high-level domain ontologies and production rules with low-level pattern recognitions and filter functions in inference processes. The approach helps to generate or check parts of building shells, technical installations, interior equipments, and exterior surroundings. The implementation is done using the programming languages Lisp and C on the basis of data steaming from management, design, and visualisation systems. The practical background is the creation of an interactive three-dimensional building information model of a university campus for various purposes.

1 Introduction

To organise education, research, administrative, operation, maintenance, design, planning, and presentation processes at a university efficiently, spatial models of a university campus play an increasing role. Examples are the use of rooms for special events, optimisation of person traffic, scheduling of courses, search and navigation tasks, distribution of media, planning of building modifications, maintenance of technical installations, or campus visualisations on websites.

The outstanding essence of a university campus as application domain is the large number of different forms of using infrastructure objects with static and dynamic features. Students, assistants, professors, teachers, researchers, lab workers, technicians, and administrators have quite different views of the common infrastructure. The infrastructure itself is very heterogeneous and frequently changing. There are many institutions with quite different organisational, technical, and scientific needs, changing research emphases, temporary courses, or short-term projects.

The nature of all these processes is three-dimensional (3D), in principle. So, it is useful and necessary to have several 3D models of the campus that represent

D. Plemenos and G. Miaoulis (Eds.): Intelligent Comp. Graphics 2011, SCI 374, pp. 61–78.
springerlink.com

different space-dependent application aspects of the statics as well as different time-dependent states of the dynamics of the reality.

Most of the available raw data of campus objects are, however, only two-dimensional (2D) drawings combined with plain textual data. In order to meet all the needs they have to be transformed into proper 3D structures.

Modern computer graphics and CAD techniques allow the creation of precise, complex, and realistic 3D models in an interactive way. But this demands extensive manual work of specialists causing high costs over an extended period of time.

These resource costs create a need to change the traditional manual approach to model generation and checking into a more and more automated one. On the other hand, the specifics of a university campus with its different aspects and high dynamics require a marginal degree of flexibility in the automation that cannot be fulfilled by a set of fixed programmes. The requirements can only be met by a flexible knowledge-based approach.

There have been several contributions to the field of 3D modelling of campus areas in the last decade. Most of them had their focus on virtual reality and visualisation topics [1]. Otherwise, many papers have dealt with artificial intelligence for 3D tasks in general [2]. The specific processing of building models with knowledge-based technologies started approximately two decades ago [3]. Since that time, several pure knowledge representation and inference methods, but also combinations of them with neural, genetic, uncertain, and probabilistic approaches have been investigated. A major problem has always been the creation of 3D models from 2D drawings and textual data.

In the present paper a proposal is developed for a consequent knowledge-based approach meeting the needs of a campus as described above by combining traditional artificial intelligence with 3D computer graphics. The approach uses knowledge in inferences to transform data of the universe of discourse.

2 Application

The application domain of this contribution is the complete infrastructure of a university campus. It comprises the categories building shell, technical installation, interior equipment, and exterior surroundings. Figure 1 shows some typical objects usually connected with a building of a campus. We see heating, sewage, and electricity systems, but also tables, shelves, blackboards, and machines in rooms of one floor of a building. On a more global level, all the buildings are located together with streets, footpaths, squares, duct alignments, and plantations on the campus of the university.

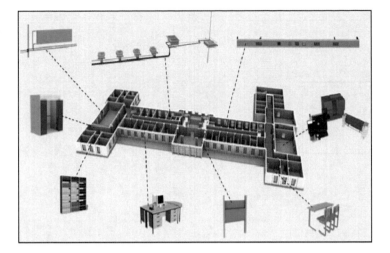

Fig. 1 Infrastructure components of a 3D campus domain

In a first step, all these objects can be created manually by 3D design tools and put together by respective configuration tools [4]. However, it emerges that there are a lot of sub-processes that are well understood so that they can be described formally, mapped into knowledge bases, computed by interpreters, and hence automated partially, in a second step. The basics of this knowledge-based technology can be found in [5]. The next three sections describe this approach shortly in a formal manner on a very global and generic level.

3 Data

The data D for analysis and synthesis processes to be automated may consist of any specific objects O of the campus. They can be divided into raw data D_R and final data D_F : $D = D_R \cup D_F$ with $D \subseteq O$. Raw data can be regarded as input and final data as output of a transformation process.

3.1 Raw Data

The amount of raw data D_R comes from 2D drawing systems, measuring devices, plans on paper, text files, or data bases. As far as format is concerned, they can be divided into 2D CAD data D_2 , image data D_I , and table data D_T . For some processes already 3D CAD data D_3 are provided as raw data:
$D_R = D_2 \cup D_I \cup D_T \cup D_3$.

Figures 2, 3, and 4 show example extracts of the used raw data: a 2D CAD drawing, a pixel image, and a relational table representing a floor plan, a laboratory equipment, and a duct system of a campus building, respectively.

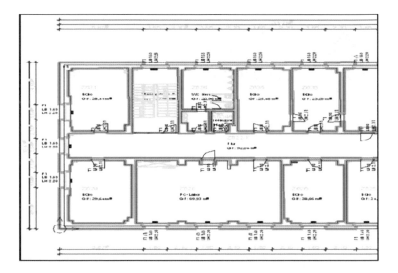

Fig. 2 2D drawing of a storey as raw data

Fig. 3 Image of a laboratory room as raw data

Endverbraucher	Bezeichnung	Raum-nummer	Zuleitung	Bezeichnung Rücklaufleitung	Ruecklauf-leitung	Bezeichnung Steigleitung
WB-Z203-1	Waschbecken	Z 203	T201	EF1-AW	A201	EF1-TKW
WB-Z204A-1	Waschbecken	Z 204B	T202	EF6-AW	A202	EF6-TKW
WB-Z208-1	Waschbecken	Z 208	T203	EF2-AW	A203	EF2-TKW
WB-Z211-1	Waschbecken	Z 211	T204	EF4-AW	A204	EF4-TKW
WB-Z215-1	Waschbecken	Z 215	T205	EF7-AW	A205	EF7-TKW
WB-Z222-1	Waschbecken	Z 222	T206	D2-AW	A206	D2-TKW
WB-Z224-1	Waschbecken	Z 224	T207	D3-AW	A207	D4-TKW
WB-Z225-1	Waschbecken	Z 225	T209	D5-AW	A209	D5-TKW
WB-Z225-2	Waschbecken	Z 225	T209	D5-AW	A209	D5-TKW

Fig. 4 Table of water ducts and devices as raw data

3.2 Final Data

The final data elements D_F are typically 3D CAD models with generated components or checked and marked subcomponents. In principle, the results of the synthesis and analysis processes may also be 2D drawings, images, or tables with textual contents. So, final data can consist, formally, of the same components as raw data: $D_F = D_2 \cup D_1 \cup D_T \cup D_3$. Figure 5 demonstrates an example of a 3D CAD model of a building shell with the building elements of several storeys. Further examples of synthesised or analysed 3D campus objects as final data are contained in figures 8, 9, 10, and 11.

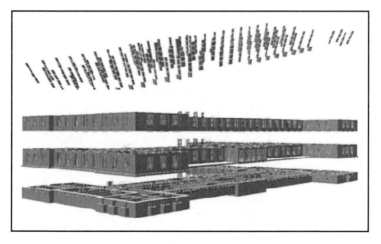

Fig. 5 3D model with building shell elements as final data

4 Knowledge

The knowledge K consists of flexible knowledge elements that can be used to formalise the contents of analysis or synthesis processes over the campus data. They are generic campus objects O originating from standards, norms, guidelines, regularities, conventions, or experiences. The centre stage is taken by a combination of high-level (semantic) domain ontologies K_O and production rules K_R with low-level (syntactic) recognition patterns K_P and functional filters K_F :

$$K = K_O \cup K_R \cup K_P \cup K_F \text{ with } K \subseteq O.$$

Ontologies are used to subdivide the entire generic campus infrastructure into smaller, manageable concepts. Production rules define dependencies according to regularities on different ontology levels. Patterns describe possible substructures of spatio-topological campus objects. Finally, filters formulate restrictions and manipulations on the patterns. All the knowledge is organised in an abstract knowledge space S with several dimensions and connections.

4.1 Knowledge Space

The knowledge space S has four main dimensions: composition S_C, generalisation S_G, idealisation S_I, and space S_S (see figure 6). Additionally, the knowledge elements are connected by associations S_A. By this, the range of validity of knowledge elements can be restricted, linked, and tuned on a global level.

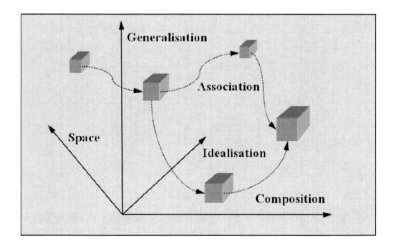

Fig. 6 Dimensions and connections of the knowledge space

Thus, there are knowledge elements for composite objects of the campus (e.g. buildings, power supplies) and for partial objects (e.g. floors, control cabinets). There are elements for generic objects (e.g. rooms, places) and for specific objects (e.g. lecture halls, parking places). There are elements for idealistic objects (e.g. topology networks, feature descriptions) and for realistic objects (e.g. pixel matrices, measurement values). The space dimension distinguishes between objects with different locations (e.g. 3D, 2D geometries). Additionally, all knowledge elements can be linked among each other by associations (e.g. contexts or activations).

The resulting knowledge space is defined by $S = (S_C, S_G, S_I, S_S)$ with a unique mapping $M : K \rightarrow S$. The association S_A is defined as a binary relation $S_A \subseteq K \times K$.

In the same way, all the data D can be included into the knowledge space and organised adequately by $M : D \rightarrow S$ and $S_A \subseteq D \times D$. Thereby, $S_G' \leq S_G''$ is true for $S' = M(D)$ and $S'' = M(K)$.

4.2 Ontologies

An ontology is understood as an interlinked set of generic semantic object classes with features and constraints in declarative form that can be used primarily to recognise or analyse respective instances (like connecting doors or stairs). Formally, an ontology $k_O \in K_O$ is a set of semantic descriptions of objects O_D connected by a set of semantic relations O_R building a network structure: $k_O \subseteq O_D \cup O_R$ with $O_R \subseteq O_D \times O_D$.

4.3 Rules

A rule defines a generic semantic mapping with condition and action parts in procedural form that interacts by a blackboard with other rules by achieving and testing object instances or features primarily for generation or synthesis purposes (like office configuration or duct connection). A rule $k_R \in K_R$ is a mapping of expressions belonging to subsets of an ontology: $k_R : P(O_D \cup O_R) \rightarrow P(O_D \cup O_R)$ where $P(X)$ stands for the set of all subsets of X.

4.4 Patterns

A pattern is a generic syntactic structure that defines a spatio-topological constellation of primitive objects in declarative form usable primarily to recognise or analyse respective instances (like lab equipment shapes or window arrangements). A pattern $k_P \in K_P$ is a spatial object structure O_S with respective topologic arrangements O_A to other objects: $k_P \subseteq O_S \cup O_A$ with $O_A \subseteq O_S \times O_S$.

4.5 *Filters*

Finally, a filter is a generic syntactic function that describes changes of patterns within the local arrangement in procedural form to perform primarily generation or synthesis tasks (like chair arrangement or room size calculation). Formally, a filter $k_F \in K_F$ is a function computing a new pattern subset from an old one:

$$k_F : P(O_S \cup O_A) \rightarrow P(O_S \cup O_A) \ .$$

5 Inference

An inference I allows the use of the knowledge to automatically transform the existing raw data into new final data whereas all kinds of knowledge are brought to bear. It comprises ontology matching, rule application, pattern matching, and filter application (see figure 7). Formally, an inference step is a mapping: $I : D \times K \rightarrow D$ with $M(D) = (S_C, S_G, S_I, S_S)$. It defines a direct derivation $D' \Rightarrow D''$.

The combination of a set of single inference steps in a dynamic chain or tree structure results in a multi-step inference process that can be controlled by different search strategies like forward, backward, depth-first, breadth-first, blind, or heuristic ones. The result is an indirect derivation $D' \Rightarrow ... \Rightarrow D''$, the transitive closure of a set of direct derivations.

In the campus domain this pertains to a complex consistency check of a duct system or to a complete generation of 3D building shells from 2D drawings, for example.

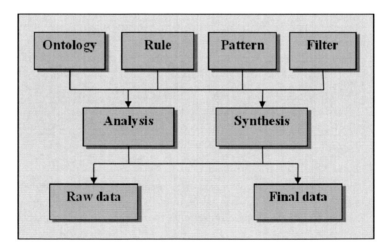

Fig. 7 Structure of the inference

5.1 Analysis

Analysis is a special inference method where idealistic objects are derived from realistic ones: $I : D' \times K \to D''$ with $S_I' \leq S_I''$. That is, meaningful, semantic objects close to ideal, mental processes are produced (for instance the assessment of utilised room capacities).

5.2 Synthesis

Synthesis handles the other direction of an inference method. It derives realistic objects from idealistic ones: $I : D' \times K \to D''$ with $S_I' \geq S_I''$. That is, structured, syntactic objects close to real, physical processes are produced (for instance the layout of positioned room equipments).

6 Transformation

In this section domain-specific knowledge-based analysis and synthesis methods are described which transform data of campus infrastructure objects on the basis of the introduced generic knowledge-based methods. They concern the four application categories building shells, technical installations, interior equipments, and exterior surroundings. The concept is open for the introduction and implementation of further similar, but also quite different methods.

6.1 Transformation of Building Shells

The automatic generation of building shells is performed by a production rule interpreter that uses pairs of conditions and actions to perform transformations from 2D drawings to 3D models [6]. Figure 8 shows an example result for a multistorey campus building with floors, ceilings, walls, pillars, bearings, roofs, doors, windows, and stairs.

The condition parts of the rules contain complex context-sensitive geometric constellations in 2D drawings, and the action parts contain pre-defined parameterised 3D building elements like cuboids or respective generation functions like extrusions. Also complex-shaped objects like stairs or roofs can be handled. The rules are organised in a spatio-topological ontology that maps possible building structures. The following Lisp expression is an extract of such a domain-specific ontology.

```
(((room part storey)(corridor part storey)
  (office sub room)(corridor sub room)
  (office connect corridor))
 (office(size small)(use personal))
 (corridor(size large)(use public)))
```

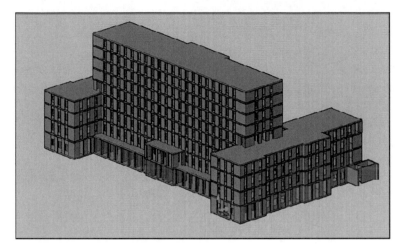

Fig. 8 Synthesised 3D building shell as final data

6.2 Transformation of Technical Installations

Technical installations are ducts for heating, water, or electricity and the respective end user devices. For correct operation it is necessary to check or generate topological and spatial features like the connectivity between inlet and outlet by chain or tree structures of ducts or the spatial relations of ducts and devices to walls, ceilings, or cavities [7]. Figure 9 contains a heating example in a building part with two heating devices and connected pairs of twig, branch, and trunk ducts.

Fig. 9 Analysed 3D technical installation as final data

The usual case is the automated consistency check. But in some cases, if only the inlet and outlet is available, production rules can be used to close the open gap by a duct chain. The input for the transformation process is given by spatial and topological duct and device arrangements, technical parameters, as well as duct ontologies and change rules. The next Lisp expression contains an example rule for the insertion of new ducts.

```
((and(inlet x)(medium x 'heat)(outlet y)
  (medium y 'heat)(bore x e)(bore y f)) >>
(and(trunkduct m)(branchduct n)(connect m n)
  (connect x m)(connect n y)(bore m e)(bore n f)))
```

6.3 Transformation of Interior Equipments

One task for the handling of interior equipments is to generate a complete 3D model of a room from a 3D room shell model, some photos taken, a textual equipment list, and a set of predefined 3D equipment objects [8]. Figure 10 contains a synthesis result with an arrangement of tables, chairs, computers, a blackboard, and a beamer.

Fig. 10 Synthesised 3D interior equipment as final data

The knowledge applied consists of an ontology describing classes of typical rooms with their equipment as well as colour, texture, and shape patterns used to find indices for objects in the images. The following Lisp expression gives an example of a pattern used to find backs of chairs in images taken by cameras.

```
((chairobj contain chairpat)
 (chairobj contact tableobj)
 (chairobj(geometry polygon)
  (size small)(convexity true))
 (chairpat(mode line)(distance 2)
  (direction 45)(colour blue)))
```

6.4   Transformation of Exterior Surroundings

The raw data for the modelling of exterior surroundings are 2D maps, aerial images, height measurements, and 3D buildings. The knowledge mainly stored in the form of filter functions can be used to partially analyse and synthesise 3D models of the spatial area of a campus and to integrate existing 3D building models. Figure 11 contains an image combined with integrated 3D building objects.

Fig. 11 Analysed 3D exterior surroundings as final data

As knowledge elements filters based on 2D patterns come into operation for the selection of relevant objects as well as for the mapping of images on the generated 3D maps with streets, places, buildings, and vegetation. Finally, all mappings have to be checked for consistency. The Lisp expression shows an example filter for the detection of streets and their correct placement.

```
((detectstreet(x)
  (if(and(equal(geometry x)'line)(<(length x)10))
    (list(name x)'class 'street)))
 (placestreet(x dist angle)
  (transl(rotate x angle)dist)))
```

7   Implementation

This section describes how the knowledge-based approach is implemented. Because of the heterogeneity and dynamics of the software system and data format situation in the CAD environment in general and the campus domain in particular, we also decided to use a heterogeneous and dynamic concept for the implementation. In this section we will have a look at the designed system architecture, the used software components, as well as the provided languages and interfaces.

7.1 System Architecture

Figure 12 contains the system architecture of the implemented knowledge-based system. It shows on the left hand side the data stream with raw data coming from the outside and final data going to the outside back. All the data concerned are stored in a relational data base system which cooperates with a file-oriented document management system.

Usually, the transformation of the data can be carried out manually via several intermediate steps. In parallel, there exists a transformation component that accompanies and supports the transformation process using different knowledge interpreters and a respective knowledge base. To handle the different data and knowledge formats properly additional interface interpreters involve language and format specifications stored in an interface base. The software system architecture is based on [9].

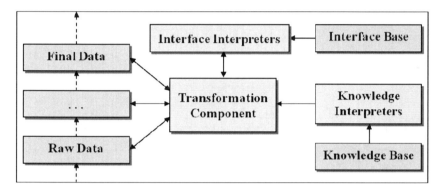

Fig. 12 System architecture

The knowledge-based system is implemented on the basis of several software components using several languages and interfaces. Those who are playing an essential role in the combination of graphics and knowledge are shown in figure 13.

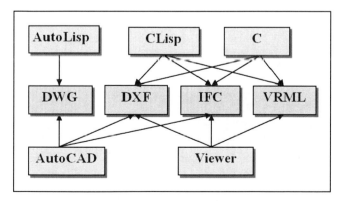

Fig. 13 Software components, languages, and interfaces

7.2 Software Components

The software developed does not build a unitary, consistent system, but a library of software components that can be applied separately or in combination on streams of raw and final data in a flexible manner.

As for the work with graphic data, the system AutoCAD plays the central role. It allows the interactive manual work, the involvement of programmes, and the data visualisation on the basis of the internal native DWG data format. For the external formats DXF, IFC, and VRML there exist several additional viewers.

Most programmes and data supporting the use of knowledge itself are written in Lisp dialects because of the capability of dynamic symbolic programming as well as the large number of interpreters and amount of experience available there.

As for knowledge processing, we use own Lisp implementations of deduction- and analogy-based interpreters. So, there are interpreters for production rules, logic clauses, frame structures, semantic networks, or similarity measurements available on the implementation level.

For programming tasks within the AutoCAD system the control language AutoLisp is used. For external formats a CLisp interpreter is the powerful basis.

Some of the low-level routine programmes, especially the filter functions, are implemented in the more efficient C language. For image pattern recognition we use additionally the HALCON graphics library and for photorealistic matters the 3DSMAX graphics and animation tool.

Aside, the data management is done by MySQL and PHP and the web presentation and access by HTML, VRML, and JavaScript.

7.3 Languages and Interfaces

The basis for handling raw and final graphic data is primarily the native format DWG. For exchange, interfacing, visualisation, and web application tasks we use the low-level graphic DXF format, and the high-level object-oriented IFC format. The web-oriented multimedia languages VRML and JavaScript are used to generate content for internet presentations and virtual reality environments. For all that, a great number of low-level filter functions have been implemented operating as generic or specific format transformations.

We don't want to miss to mention that there are a lot of small, but annoying problems concerning the transformation of data formats along to the documentations of software systems and standards.

The graphics-oriented formats are supplemented by standard data formats for pixel images, relational tables, and plain texts.

The syntax and semantics of knowledge representation languages depend on the knowledge-based interpreters that are implemented in the system. For convenience, sets of Lisp expressions or C strings are used. The knowledge is stored in text files which are easy to handle.

The specially written interface components make it possible to read, traverse, and write all data and knowledge structures of the different languages, respectively formats in connection with different application-dependent query, update, and delete functions.

8 Evaluation and Test

Generally, the knowledge-based components are part of a management and web application system for campus infrastructure elements with a data base and document management system as core that store and provide all the data on different levels of abstraction. Within this framework all the described components have been implemented, tested, and evaluated. The short-term result is a set of infrastructure objects with partially automatically and manually generated and checked components.

The long-term result is given by the knowledge base and the interpreters that have been and will be developed step by step. The degree of the tests of single components is quite different. The generation of building shells was successfully done for all buildings of the campus. The generation of the interior equipment has been carried out for selected public rooms with certain well-known generic criteria. Automated consistency checks have been applied permanently. They support the manual design work, especially for the technical installations. The handling of exterior surroundings is currently focussed on the integration of 2D maps, aerial photos, and 3D buildings and still in process.

An important aspect of the evaluation and test of the system components concerns visualisation tasks. Several transformation functions have been used to prepare 3D models for photorealistic image or video presentations (for instance in [10]). Figure 14 shows an image of a lecture hall rendered in the context of specific light conditions. Figure 15 shows a stairway connecting two storeys.

Fig. 14 Rendered image of a 3D lecture hall

Fig. 15 Rendered image of a 3D stairway

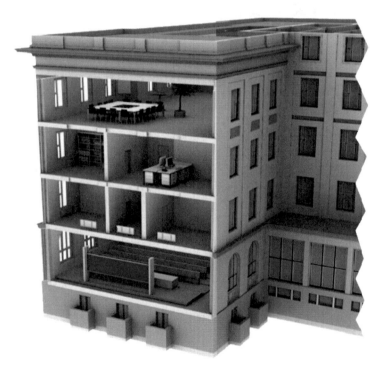

Fig. 16 Web-based virtual reality presentation of a campus infrastructure object

On the other hand, interactive navigation environments of the entire campus in-frastructure or parts of it have been created as virtual environments. The visualisation of such environments can be done by traditional 2D screens, by auto-stereoscopic displays, or projections on power walls. Figure 16 gives an impression of such a virtual real world that can be accessed on a local computer or via the internet.

Both kinds of visualisations give us permanently a valuable feedback about the correctness of the manual work and the implemented knowledge-based components. Practical considerations and evaluations from the point of view of civil engineering and architecture have also been made and can be found in [11].

Summarizing the work done we can assess that the data and knowledge acquisition process is finished. The data base is complete and its management is working well. The knowledge interpreters have been implemented and tested in principle with a selectively filled knowledge base. On this basis we have proved the feasibility, flexibility, usability of the knowledge-based approach, but the quality and quantity of the results depend mainly on the contents of the knowledge base that is still under development.

9 Conclusion

The paper described a consequent knowledge-based approach to support the generation and checking of 3D campus infrastructure models according to the requirements of the application domain. The implementation began after a manual creation of large parts of the campus infrastructure as a first contribution to a long-term process of sustainable modelling and permanent adaptation. It can be regarded as a systematic and unified approach to a heterogeneous and dynamic application domain.

Until now, a comprehensive 3D model of all major objects of a university campus has been built up. Within this practical context a lot of basic technologies have been designed, developed, and tested. This process will be continued. The current stage can be characterised as a manual human-driven modelling with a selected knowledge-based support to reach a static 3D campus infrastructure model.

The future task will be to improve and accomplish these methods and put them together to an integrated, but nevertheless simple, efficient, and flexible system. In the long-term view, this will result in a smooth change from the current data base system to a future knowledge base system for campus infrastructure models. Then, a possible future stage will be characterised as an automated machine-driven modelling by an extensive knowledge-based support to reach a dynamic 4D campus infrastructure model.

Acknowledgments. The contents of the paper is a result of the project "Virtual three-dimensional campus infrastructure model (V3CIM)" supported by the Saxon State Ministry of Science and Art from 2009 to 2010. The project joined scientists of the faculties of information technology / mathematics, civil engineering / architecture, mechanical engineering / process engineering, and surveying / cartography. It is currently continued by a project with the same funding under the title "Sustainable Campus" with the focus on the temporal and long-term aspects of campus infrastructure modelling. Our thanks go to the members of the project group for providing the professional context of the work. We also have to thank all the students who have created building-relevant models or implemented program prototypes in the scope of CAD, CG, and VR courses, project seminars, workshops, and final year projects, or as student assistants.

References

1. Parasolova-Forland, E., Sourin, A., Sourina, O.: Cybercampuses: Design Issues and Future Directions. The Visual Computer 22(12) (2006)
2. Kalogerakis, E., Christodoulakis, S., Moumoutzis, N.: Coupling Ontologies with Graphics Content for Knowledge Driven Visualization. IEEE Virtual Reality (2006)
3. Bakhtari, S., Bartsch-Spoerl, B., Oertel, W.: DOM-ARCADE: Assistance Services for Construction, Evaluation, and Adaptation of Design Layouts. In: Gero, J., Sudweeks, F. (eds.) Artificial Intelligence in Design 1996. Kluwer Academic Publishers, Dordrecht (1996)
4. Kantardshieffa, H.: Technology of 3D Data Transformation for Interactive Virtual Building Infrastructure Models. In: Scharff, P. (ed.) International Scientific Colloquium. TU Ilmenau (2010)
5. Oertel, W.: Systementwicklung auf der Basis einer dynamischen Wissensorganisation. Dresden University Press, Muenchen (1999)
6. Bauer, R., Biel, M., Porsch, F., Schoenfeld, I., Wolff, S.: Erzeugung von Gebaeude- und Navigationsmodellen aus Zeichnungen. Projektseminar VRIE, HTW Dresden, Informatik/Mathematik (2009)
7. Gaunitz, R., Noubi, M., Schrage, M., Wenk, T., Zerjatke, C.: Datenbankgestuetzte Verwaltung technischer Gebaeudeausruestungen. Workshop VRIE, HTW Dresden, Informatik/ Mathematik (2010)
8. Alkhouri, R., Haase, R., Klawikowski, N.: Erzeugung von Raum-Modellen aus Bildern. Projektseminar VRIE, HTW Dresden, Informatik/Mathematik (2010)
9. Kantardshieffa, H., Oertel, W.: Software Concept of a Three-Dimensional Campus Infrastructure Model. In: GFaI: 3D-NordOst 2009, GfaI, Berlin (2009)
10. Vierheller, T.: Unterstützung informationeller Prozesse durch 3D und 4D Visualisierungen im Rahmen eines virtuellen dreidimensionalen Campus-Infrastrukturmodells. Diplomarbeit, Bauingenieurwesen/Architektur, HTW Dresden (2009)
11. Just, M., Kunze, U.: Intelligente 3D-Daten: V3CIM - Das virtuelle dreidimensionale Campus-Infrastrukturmodell der HTW Dresden. In: Wissend., vol. 19(1). Magazin der HTW Dresden, HTW Dresden (2011)

On the Evaluation of the Combined Role of Virtual Reality Games and Animated Agents in Edutainment

Maria Virvou

Department of Informatics,
University of Piraeus,
80 Karaoli & Dimitriou St., Piraeus 18534
mvirvou@unipi.gr

Abstract. Edutainment is a form of education combined with entertainment and vice versa. Virtual reality games and animated agents can provide high quality graphics that may play an important role to serve the aims of edutainment. However, since in edutainment the main aim of the software developed lies precisely on the combination of education and entertainment, there have to be appropriate evaluation methods that evaluate both these factors and assure that their combination is worth the effort and the development cost. So far, in the literature there have been evaluation frameworks that address each factor separately, but there is a shortage of frameworks on the combined role of education and entertainment of Virtual Reality games and animated agents. In this paper, I will survey current trends concerning the evaluation of educational software and Virtual Reality games and animated agents and will speculate about future challenges and design enhancements.

Keywords: virtual reality games, educational games, edutainment, animated agents, evaluation.

1 Introduction

Edutainment software aims at educating users while they are being entertained. Thus, edutainment software is developed in order to serve the combined role of an educator and an entertainer. On one hand, educational software aims at helping students to learn about a domain that is being taught. Learning may provide a great joy in itself to the learner, since his/her curiosity about a new unexplored topic may be satisfied through tutoring. Despite this joy, learning also poses a great cognitive burden on students, who have to make great efforts to understand and learn the new unknown material. This burden may be very tiring and repelling, especially if the presentation of the teaching material is not attractive. Presentation of the teaching material plays an important role in stimulating students and keeping them alert. On the other hand, there is a whole imposing culture of computer games that provide excitement and immersion to children and adolescents as has already been acknowledged by many researchers. Indeed, many researchers

D. Plemenos and G. Miaoulis (Eds.): Intelligent Comp. Graphics 2011, SCI 374, pp. 79–95.
springerlink.com
© Springer-Verlag Berlin Heidelberg 2012

have conducted research studies where they show that computer game playing is by far the most popular activity on the computer for children and adolescents (e.g. [4], [6], [12] and [14]). One important asset of computer games is based on advanced graphics, such as virtual reality and animated agents which are particularly attractive to computer users. In this respect the combination of virtual reality games and animated agents with education seems very promising for potential learners.

However, despite the nice features of virtual reality games and animated agents, the effectiveness of edutainment software that relies on them has to be evaluated. There are many questions to be answered with respect to effectiveness of this kind of software: Is it educationally beneficial? If not, what is the point of incorporating high quality graphics and making the educational application heavy and more difficult to design?

Is it sufficiently entertaining? If not, what is the point to design an edutainment application if this does not increase the students' motivation?

Besides the above fundamental questions, there are other questions that arise from the use of complex graphics in edutainment contexts. For example, how easy is it for a student-player to navigate inside the application? Another example is how easy it is for a student, who is a naïve player, to play the educational game so that s/he may gain the benefit of learning. Is the gaming environment with animated agents distractive? Is it irritating? Does the educational context render the game boring? Does the gaming environment render the educational aims difficult to attain?

Indeed, there have been researchers that seem quite skeptical about edutainment in this respect. For example, Brody [2] made a criticism on video games that the marriage of education and game-like entertainment has produced some not-very-educational games and some not very-entertaining learning activities. This criticism may hold for VR-games with animated agents. Moreover, Yacci and colleagues [28] , argue that edutainment environments which include educational games, demand a certain amount of effort and learning that is not related to the instructional goals of the school lesson that is taught; such irrelevant learning concerns the plot or mission of the game and the "legal" movements and actions that a player can make while "inside" the game. Thus, they conclude that there is a very important question associated with educational games: how much student effort is an expense when engaging in edutainment. Indeed, despite the great popularity of games there are still a lot of children that do not know how to play and perhaps need a lot of effort to learn how to play on top of the cognitive effort that they need to make in order to learn the educational content.

In view of the above, the development of edutainment software that is based on high quality entertainment graphics such a VR-games and animated agents, has to be evaluated on many different aspects, to make sure that it serves well the purposes of its creation. However, in the literature there is a shortage of evaluation guidelines or evaluation frameworks dedicated to edutainment software and especially VR-games and animated agents. There exist evaluation guidelines and frameworks for educational software alone but are not directly appropriate for the evaluation of the combined functionality of education and entertainment.

The main body of this paper is organized as follows: In Section 2 some existing evaluation frameworks (mainly concerning educational software and only one concerning edutainment) are surveyed and discussed. In Section 2, I will discuss the role of underlying reasoning and intelligence for VR-educational games and animated agents as an important factor of successful development of edutainment software of this kind. Then, I will survey previous extended research on empirical evaluations that has been performed within our own research lab and the main conclusions that were drawn from these evaluations are highlighted and discussed. In Section 3, I will speculate on evaluation guidelines for edutainment software. Finally in Section 4 the conclusions of this paper are presented and also future challenges and design enhancements are suggested.

2   The Role of Underlying Reasoning and Intelligence for the Success of Edutainment VR-Games and Animated Agents

Edutainment VR-games and animated agents provide more human-like environments than the more conventional user interfaces. In this sense, edutainment VR-games and animated agents generate high expectations to users who may be disappointed if the underlying reasoning mechanisms of the systems created, fall short of their expectations. Indeed, animated agents can render an application more user friendly and pleasant but on the other hand may cause distraction, disappointment and annoyance. For example, Swartz (2003) argues that the well-known Microsoft agent "Clippy the Paperclip" (the default character, referred to in Microsoft Office itself as "Clippit"), seemed to have attracted widespread negative opinion until it was finally withdrrawn by Microsoft after the release of Windows XP.

Moreover, in edutainment, users may have very different profiles and skills. This is so, because there may be users who are mainly attracted to the entertainment part (and thus they might perform well at game playing) while others may be better at the educational part. Adaptability may look like a solution to this problem. Adaptability is the functionality of a software application that allows users to adapt the user interface of the application to their needs and preferences. Adaptability refers to the user-initiated modification. On the other hand, there is also another type of functionality referring to system-initiated dynamic modifications of user interfaces to serve the personal needs and preferences of users. This functionality is known as adaptivity. In any case adaptable software always incorporates some kind of user modelling techniques. However, in the case of adaptability, which refers to the user's ability to adjust the form of input and output, this customisation may be very limited. In many cases, the user is only allowed to adjust the position of soft buttons on the screen or redefine command names.

As such, adaptability is not flexible enough to provide the desired effects to users. Indeed, in the case of adaptability, users have to know how to customise software to their particular needs. If the software is complex, as is the case in edutainment software, then a user may not even know what his/her needs are, let alone how to customise the software themselves. On the other hand, adaptivity provides automatic customisation of software to the users' needs based on more sophisticated user modelling techniques. It may also provide intelligent help, which is more flexible than traditional on-line help.

Thus, the incorporation of underlying reasoning mechanisms that rely on intelligent techniques would provide credit to the application if they are designed correctly. However, this is also an issue to evaluate.

3   Existing Evaluation Frameworks and Guidelines

The issue of evaluating educational software has been dealt by many researchers in the previous years and has been related to evaluating usability of interactive software. As in edutainment, one important factor is educational software, in Section 3.1, I present a few approaches concerning evaluation frameworks and guidelines. However, as also pointed out by other researchers, the evaluation of educational games has largely been based on leisure-based games only, despite the obvious difference between leisure-time games and educational games [5], [8], [16]. Indeed, edutainment software evaluation has not been explored adequately yet. In Section 3.2, I present one existing evaluation framework and evaluation comments that exist in the relevant literature.

3.1   *Evaluation for Educational Software*

Several frameworks have been developed concerning the synergy of computers and education in the form of Computer Assisted Learning (CAL). Some of them are the following:

The JIGSAW model [17] addresses the problem of integrating both usability and learning issues in the evaluation of educational software. In the JIGSAW model, the evaluation is performed in three levels. In Level 1, the subtasks of the learning and operational tasks are considered independently of each other. As we move from Level 1 to Level 2, integration within the learning and the operational tasks is considered. At Level 3 integration between the learning and operational tasks is considered.

Similarly to the JIGSAW model, the set of 'learning with software' heuristics [18], also address the problem of integrating both usability and learning issues in the evaluation of educational software. The set of 'learning with software' heuristics are based on the "usability heuristics" specified by Nielsen [15], so as to relate them to socio-constructivist criteria for learning. These heuristics include needs such as to consider appropriate levels of learner control, the prevention of peripheral cognitive errors and the creation of strategies for the cognitive error recognition, diagnosis and recovery cycle.

The CIAO! framework [9] outlines three dimensions to evaluate: (i) context; (ii) interactions; and (iii) attitudes and outcomes. The context refers mainly to the underlying rationale for the development and use of CAL; this is concerned with the reason why the CAL project was developed in the first place. Interactions with the software show information about the students' learning processes; such interactions can provide protocol data for later analysis. Finally, in outcomes, pre and post-achievement tests, interviews and questionnaires with students and tutors show how well CAL has attained its aims. An example of use of the CIAO! Framework, that we performed in our research lab was in the context of the formative evaluation of an Intelligent Tutoring System for the Passive Voice of the English language [27].

The Flashlight framework [7] presents a similar approach to CIAO! Framework. Its aim is to examine the relationship between three factors: A technology, the activity for which it is used and the educational outcome.

All of the above frameworks indicate the importance of using questionnaires, interviews, collection of protocols and experiments in order to find out what the educational effect has been for the students and they argue that usability issues are important within these processes. These frameworks only cope with the two main aspects of the educational software, meaning the process of learning and the process of using a computer to do so. However, in the case of edutainment there are four aspects involved: learning, computer use, pedagogy and entertainment and often these aspects compete with one another. Thus evaluation frameworks for edutainment have to be more complex.

3.2 Evaluation of Edutainment Software

As has already been pointed out, there is a shortage of evaluation frameworks for edutainment. One existing framework is the one created by De Freitas and Oliver [5]. This is a four-dimensional framework for helping tutors to evaluate the potential of using games and simulation-based learning. The first dimension focuses upon the particular context where play/learning takes place (classroom based, outdoors, access to equipment, technical support etc.). The second dimension focuses on the learner (learner profile, pathways, learning background, group profile etc.). The third dimension focuses on the internal representation world (level of fidelity, interactivity, immersion etc.) and the fourth dimension focuses on pedagogic considerations (learning models used, approaches taken etc.) This framework has focused primarily on game based learning and simulations. As such, it does not address adequately the issue of entertainment versus learning which is also quite important as in many cases the educational aims of edutainment may undermine entertainment and vice versa. Moreover, it does not distinguish clearly between the usability concerning the computer use, usability concerning the educational part of the software and usability concerning the entertainment part.

4   Empirical Evaluations Concerning Educational Software and Edutainment VR-Games and Animated Agents

In this Section, I will survey some research projects that have been carried out in our research lab in the past concerning Virtual Reality edutainment software and educational software that incorporates animated agents and I will highlight the main conclusions that were drawn form empirical studies performed. These conclusions lead to important guidelines for evaluations on edutainment software of this kind.

4.1   The Case of Educational VR Games

In our research lab, we have created several VR-games and authoring tools generating edutainment VR games that incorporate intelligence. One VR- game is called VR-ENGAGE [21], [24], [25] and teaches students geography. A later VR-game is called VIRGE [11], [22] and teaches students English as a second language over the Web. Moreover, two authoring tools have been created, so that multiple games may be generated in a cost-effective way. The first authoring tool is called VR-INTEGATE [26] and allows author-instructors to author their own educational VR-games on the domain that they wish. The other authoring tool is called VR-MultiAuthor [23] and allows multiple authors to author their own educational games which can operate in an integrated environment of educational games in multiple domains (e.g geography, physics, mathematics) and share the same VR-game platform and a uniform student model for each student that uses all of the VR-games.

In VR-ENGAGE [25] the environment of the game is similar to that of many popular adventure games which have many virtual theme worlds with castles and enemies that the player has to navigate through and reach the end of the level. In particular, it is quite similar to the commercial game DOOM [10]. The goal of a player is to navigate through a virtual world and find the book of wisdom. The total score is the sum of the points that the player has obtained by answering questions. During the game the player may come across certain objects and animated agents. These objects and animated agents ask questions, give hints to students or guide them to tutoring places. Figure 1 illustrates screenshots of VR-ENGAGE.

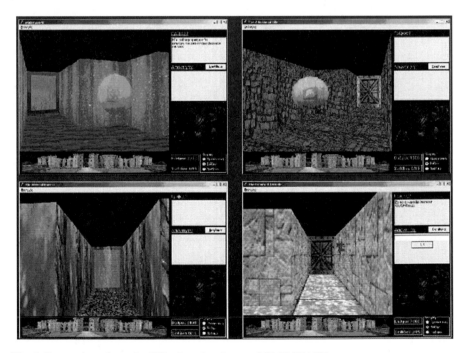

Fig. 1 Screenshots from the VR-educational game VR-ENGAGE

Fig. 2 Animated Agents from the VR-educational game VIRGE

In VR-ENGAGE there is also an underlying user modeling module that performs cognitive student modeling. This student modeling module is divided into two main functionalities. One functionality concerns student error diagnosis. Error diagnosis is performed by the virtual enemy of the student-player, the dragon who guards passages and doors. If a student answers incorrectly to a question asked by the dragon about the domain being taught then the dragon performs error diagnosis and if the error is not very serious it allows the user to go on in the virtual world.

In VIRGE, there is more than one agent. For example, there is the animated agent featuring an angel that performs emotion recognition with respect to the users' feelings and tries to give advice and empathy to the student-player in difficult situations. In the case of VIRGE, the underlying user modeling module performs emotion recognition taking into account both the circumstances of the game as well as the educational application. Animated agents that take part in the educational game can be seen in Figure 2 which illustrates two different screenshots of VIRGE.

VR-MultiAuthor is a knowledge-based authoring tool for Intelligent Tutoring Systems that operate as virtual reality computer games, and focused on its player modelling capabilities. VR-MultiAuthor models two broad categories of user characteristics: the domain dependent and the domain independent. Domain dependent features mainly concern the particular domain being taught. On the other hand, domain independent features concern the player's behaviour in the game irrespective of the content of questions being asked to him or her. In this way, domain-independent features may be used for modelling an individual student across many application domains and provide clear separation of the user's skills as a VR-game player from his /her competence level in the domain being taught.

In all of the above games and authoring tools certain features of students were identified that concerned their game playing skill:

1. Virtual Reality User Interface Acquaintance

In this feature the system measures whether the player knows the functionality of virtual reality user interface features that are important for the game, such as the "Inventory", the "Tutor" etc. For example, there may be a player in an occasion where s/he should have used a key inside the player's inventory, in order to continue playing but the user is not responding. Then this player is probably ignorant of his/her inventory's usage.

2. Navigational Effort

Not all users know how to play a 3D Virtual Reality Game. This feature measures how well the user can navigate through the Virtual World. For example a character may be bumping onto a wall (or other virtual items), instead of getting to the door, or may even rotate around the same position without proceeding to the next places.

3. VR Environment Distractions

There are many cases when the Virtual Environment draws the player's attention so much that s/he may miss the main point of the educational game (which is learning a specific subject). This is the case in situations such as when a player finds a door, does not answer the riddle, goes back to the previous encountered tutor (angel), reads the hint, goes to the door, does not answer the riddle, goes again to the tutor etc. This behaviour shows that from the tutor to the door the player may have been so distracted that s/he forgot the hint.

Domain dependent information about a player is kept separately in the corresponding domain application. On the other hand, domain independent information about each student is kept in the integrated learning environment so that it may be used and updated by all available domain applications. The domain independent features are illustrated in Figure 3.

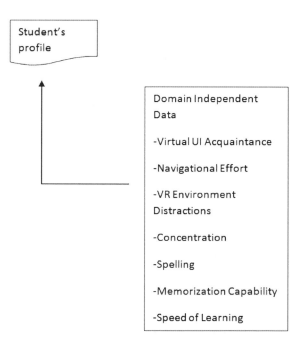

Student's profile

Domain Independent Data

-Virtual UI Acquaintance

-Navigational Effort

-VR Environment Distractions

-Concentration

-Spelling

-Memorization Capability

-Speed of Learning

Fig. 3. Domain-independent parts of player models, irrespective of the domain being taught

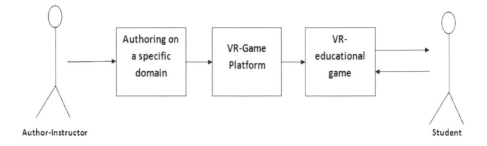

Fig. 4 Architecture of the authoring tool VR-INTEGATE

VR-INTEGATE provides an authoring environment to instructors who wish to create Intelligent Tutoring Systems (ITSs) that operate through a virtual reality game. The concept of the game is used so that the resulting ITSs may become more motivating and engaging. Moreover the ITSs are able to provide diagnostic reasoning concerning the students' answers to questions about the domain being taught. The initial input to the authoring tool is given by a human tutor who is acting as an author. Then the authoring tool provides the VR-game platform and the underlying reasoning and constructs the VR-educational game to be used by students. The architecture of VR-INTEGATE is illustrated in Figue 4.

All of the above games and tools were evaluated. In this paper I will describe shortly the evaluation of VR-ENGAGE. The evaluation of VR-ENGAGE consisted of two phases. On one phase there was an evaluation concerning the educational effectiveness of VR-ENGAGE as compared to a similar educational system that was not game-based [24]. The results of the evaluation showed that educational virtual reality games can be very motivating while retaining or even improving the educational effects on students. Moreover, one important finding of the study was that the educational effectiveness of the game was particularly high for students who used to have poor performance in the domain taught prior to their learning experience with the game.

On the second phase there were evaluation experiments concerning usability and likeability [21]. If the games acquire an educational content they may lose the attractiveness and appeal that they have on experienced users who are familiar with commercial games. On the other hand, usability may be a problem for novice game players and inexperienced computer users. The evaluation experiments that were conducted, involved 50 students from schools who were asked to play the game both in and out of the classroom environment. These students included three categories of game-players in terms of their level of game-playing expertise: novice, intermediate and expert game players. The likeability of the educational game was also evaluated by its comparison with educational software of no gaming environment and with commercial games of no educational content in terms of the students' likeness of them. The evaluation results showed that the game was indeed usable and likeable but there was scope for usability and likeability improvement so that the educational benefits may be maximised for all categories of students.

One important issue in the evaluation experiments concerning VR-ENGAGE was the fact that the game users were distinguished among three categories of students ranging from students who typically had a good performance at school lessons to students who typically had a poor performance. Similarly, usability and likeability was evaluated taking into account three categories of game players ranging from experience game players to novice ones. Indeed, this was useful as the results were different depending on the respective categories the users belonged to.

4.2 The Case of Animated Agents

In our research lab we have also created many CAL projects that incorporate intelligent animated agents. One such application is an authoring tool, called WEAR, that creates Intelligent Tutoring Systems that use animated agents [13] to communicate orally the messages of the system to the student. Another application is an authoring tool that allows authors to create emotion sensitive teaching personalities that are embodied in animated agents [1], [20]. For example, Figure 5 illustrates a screenshot of the Medical Tutor where the tutor-animated agent can be seen at the bottom of the screenshot and Figure 6 illustrates other animated agents that were incorporated and embodied the reasoning of emotion generation by the system.

In the case of WEAR, we conducted evaluation experiments concerning the likeability and educational effectiveness of an animated agent that had been embodied in the interface of an Intelligent Tutoring System (ITS) [13]. This agent was responsible for guiding the student to the learning environment and providing feedback messages in an orally speaking mode. The agent was evaluated during an experiment in terms of the effect that it could have on students' learning, behaviour and experience. The participants in the experiment were divided into two groups: half of them worked with a version of the ITS which embodied the agent and the rest worked with an agent-less version. The results from this study confirmed the hypothesis that an animated agent incorporated in an ITS could enhance students' learning experience. On the other hand, the hypothesis that the presence of the agent improved short-term learning effects was rejected based on our data. The presence of the interface agent did not manage to improve significantly the short-term learning outcomes. Furthermore, the students' attentiveness to the system was not promoted by the agent.

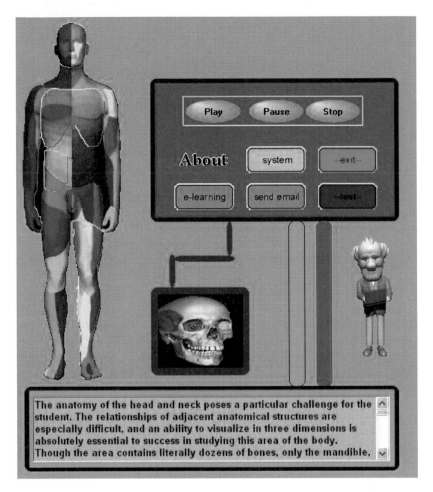

Fig. 5 Screenshot from the Medical Tutor platform

Fig. 6 Animated agents in different situations

This may not seem very encouraging at a first glance. However, the use of animated agents did promote the learning process significantly since it was confirmed that it increased the students' motivation. This may show positive results in learning outcomes in the long run, since motivated students may dedicate more time on the edutainment software and get a better consolidation of the material taught.

5 Evaluation Guidelines Concerning Edutainment VR-Games and Animated Agents

In view of the above, an evaluation of edutainment VR-games and animated agents should be oriented around the 3 roles of the respective human users, namely student, player and computer user. Moreover it should address separately each one of the fundamental functionalities, namely education, entertainment and usability, and their synergy, as illustrated in Figure 7. In particular, usability is an issue underlying both educational aims and entertainment aims.

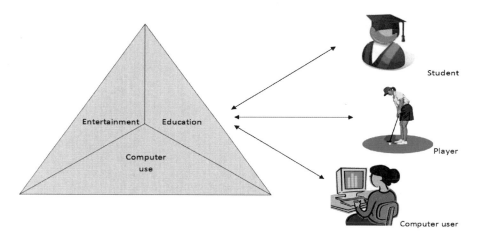

Fig. 7 Functionalities and human user roles in edutainment

It is useful in these experiments to distinguish among the proficiency level of end users with respect to each of their three roles. For example, edutainment software may have a different impact on a student who has typically a good performance at school lessons, but has intermediate experience in the use of computers and no experience at all in VR-game playing (as illustrated in the example of Table 1) from another student who has typically a poor performance at school lessons but is an expert in the computer use and an expert VR-game player. Ideally, these categories human roles and levels of proficiency of each human user with respect to each of these roles should be taken into account in the design of edutainment software by incorporating adaptability features or better adaptivity features which rely on intelligent techniques. However, if adaptivity is incorporated then again this adaptivity has to be edvaluated.

Table 1 An example of a target user classification

Proficiency level / Human roles	Advanced	Intermediate	Novice
Student	x		
Player			x
Computer User		x	

Evaluations should be conducted at three levels. On the first two levels education effects and entertainment effects should be evaluated separately and on the third level, the synergy between them and the tension between them has to be evaluated.

Education:

1. Pedagogy
2. Underlying Reasoning
3. Educational effectiveness
4. Usability

Entertainment:

1. Amusement
2. Distractions
3. Usability

Education vs Entertainment:

1. Educational content vs entertainment effects
2. Entertainment content vs social ethics within pedagogy aims
3. Entertainment complication vs comprehension of educational content

Consequently, empirical evaluation has to be performed concerning likeability of the edutainment software, usability, educational aims and pedagogic aims. Evaluation concerning each of these aspects should also evaluate the degree of personalisation of the software ranging from adaptability (at the lowest point of personalization) to adaptivity (at the highest point of personalization), as illustrated in Figure 8.

Empirical evaluation refers to the appraisal of a theory by observations in experiments [3]. Since there is a tension between the above four aims, comparative studies are recommended. For example, the comparison of educational software that contains the entertainment aspect with educational software that does not contain it.

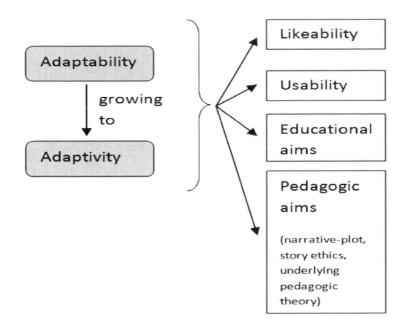

Fig. 8 Evaluation aspects of edutainment

6 Conclusions and Speculations for Future Research

In this paper we argued that edutainment VR-games and animated agents have to be evaluated in terms of likeability, usability, educational aims and pedagogic aims. One important issue of evaluations concerns the three roles of human users, who in the case of edutainment are students, players and computer users since edutainment is implemented as a computer program. Edutainment has to have the ability to distinguish between a player's ability to play the game itself and a student's level of knowledge in the particular domain being taught. This ability should be incorporated in the design so that the edutainment application becomes adaptive to the user. Players who are not familiar with the user interfaces of games should be given extra help in this respect and their level of domain knowledge should not be underestimated. However, if adaptivity and intelligent techniques are involved in the design, they should also be evaluated.

We strongly believe that intelligent techniques and VR-edutainment games and animated agents have the ability to improve many aspects of education by motivating students and engaging them in the learning process. However, there is still a lot of research needed for acquiring evaluation frameworks that will take into account all aspects of such sophisticated software.

References

1. Alepis, E., Virvou, M.: Automatic generation of emotions in tutoring agents for affective e-learning in medical education. Expert Systems with Applications 38(8), 9840–9847 (2011)
2. Brody, H.: Video games that teach? Technology Review 96(8), 51–57 (1993)
3. Chin, D.: Empirical Evaluation of user models and user adapted systems. User Modeling and User Adapted Interaction 11(2001), 181–194 (2001)
4. Cunningham, H.: Gender and computer games. Media Education Journal 17 (1994)
5. De Freitas, S., Oliver, M.: How can exploratory learning with games and simulations within the curriculum be most effectively evaluated? Computers & Education 46(2006), 249–264 (2006)
6. Downes, T.: Playing with computer technologies in the home. Education and Information Technologies 4, 65–79 (1999)
7. Ehrmann, S.: Studying teaching, learning and technology: a tool kit from the Flashlight programme. Active Learning 9, 36–39 (1999)
8. Gee, J.: What video games have to teach us about learning and literacy. Palgrave Macmillan, New York (2003)
9. Jones, A., Scanlon, E., Tosunoglu, C., Morris, E., Ross, S., Butcher, P., Greenberg, J.: Contexts for Evaluating Educational Software. Interacting with Computers 11(5), 499–516 (1999)
10. ID-software (1993)
11. Katsionis, G., Virvou, M.: Personalised e-learning through an Educational Virtual Reality Game using Web Services. Multimedia Tools and Applications 39(1), 47–71 (2008)
12. Kirkman, C.: Computer experience and attitudes of 12-year-old students: implications for the U.K. national curriculum. Journal of Computer Assisted Learning 9, 51–62 (1993)
13. Moundridou, M., Virvou, M.: Evaluating the persona effect of an interface agent in an intelligent tutoring system. Journal of Computer Assisted Learning 18(3), 253–261 (2002)
14. Mumtaz, S.: Children's enjoyment and perception of computer use in the home and the school. Computers and Education 36, 347–362 (2001)
15. Nielsen, J.: Usability Inspection Methods. John Wiley, New York (1994)
16. Prensky, M.: Digital games-based learning. McGraw Hill, New York (2001)
17. Squires, D., Preece, J.: Usability and Learning: Evaluating the Potential of Educational Software. Computers and Education 27(1), 15–22 (1996)
18. Squires, D., Preece, J.: Predicting Quality in Educational Software: Evaluating for learning, usability and the synergy between them. Interacting with Computers 11(5), 467–483 (1999)
19. Swartz, L.: Why people hate the paperclip: labels, appearance, behavior and social responses to user interface agents. B.Sc. Thesis. Stanford University (2003)
20. Virvou, M., Alepis, E.: Creating tutoring characters through a Web-based authoring tool for educational software. In: Proceedings of the 2003 IEEE International Conference on Systems, Man and Cybernetics, vol. 5, pp. 4884–4889 (2003)
21. Virvou, M., Katsionis, G.: On the usability and likeability of virtual reality games for education: The case of VR-ENGAGE. Computers & Education 50(1), 154–178 (2008)

22. Virvou, M., Katsionis, G.: VIRGE: Tutoring English over the Web through a Game. In: IEEE International Conference on Advanced Learning Technologies 2003, Athens, Greece, July 9-11, p. 469 (2003)
23. Virvou, M., Katsionis, G., Manos, K.: On the motivation and attractiveness scope of the virtual reality user interface of an educational game. In: Bubak, M., van Albada, G.D., Sloot, P.M.A., Dongarra, J. (eds.) ICCS 2004. LNCS, vol. 3038, pp. 962–969. Springer, Heidelberg (2004)
24. Virvou, M., Katsionis, G., Manos, K.: Combining software games with education: Evaluation of its educational effectiveness. Educational Technology & Society, Journal of International Forum of Educational Technology & Society and IEEE Learning Technology Task Force 8(2), 54–65 (2005)
25. Virvou, M., Manos, C., Katsionis, G., Tourtoglou, K.: VR-ENGAGE: A Virtual Reality Educational Game that Incorporates Intelligence. In: IEEE International Conference on Advanced Learning Technologies, Kazan, Russia, September 16-19 (2002a)
26. Virvou, M., Manos, C., Katsionis, G., Tourtoglou: Incorporating the Culture of Virtual Reality Games into Educational Software via an Authoring Tool. In: IEEE International Conference on Systems Man and Cybernetics 2002 (SMC 2002), Hammamet, Tunisia, vol. 2, pp. 326–331 (2002b)
27. Virvou, M., Maras, D., Tsiriga, V.: Evaluation of an ITS for the Passive Voice of the English Language Using the CIAO! Framework. In: Proceedings of ED-MEDIA 2000, World Conferences on Educational Multimedia, Hypermedia and Educational Telecommunications, Montreal, Canada, pp. 1722–1723 (2000)
28. Yacci, M., Haake, A., Rozanski, E.: Operations and Strategy Learning in Edutainment Interfaces. In: Proceedings of ED-MEDIA World Conference on Educational Multimedia, Hypermedia and Telecommunications 2004, pp. 4416–4421 (2004)

Serious Games for Cultural Applications

Anastasios Doulamis [1], Fotis Liarokapis[2], Panagiotis Petridis[3],
and Georgios Miaoulis[4]

[1] Technical University of Crete,
   Decision Support Lab.
   Chania, 73100, Crete, Greece
   Tel.: + 30 28210 37430
   adoulam@ergasya.tuc.gr
[2] Coventry University
   Coventry, United Kingdom, UK
   Interactive Worlds Applied Research Group
   Tel.: +44 (0)24 7688 7631
   F.Liarokapis@coventry.ac.uk
[3] Serious Games Institute (SGI)
   Coventry Innovation Village
   Coventry University Technology Park
   Cheetah Road, Coventry, West Midlands, CV1 2TL
   Coventry, United Kingdom, UK
   PPetridis@cad.coventry.ac.uk
[4] Technological Education Institute of Athens
   Department of Informatics
   Ag.Spyridonos St., 122 10 Egaleo, Greece
   ndoulam@cs.ntua.gr, idrag@teiath.gr, gmiaoul@teiath.gr

Abstract. This paper presents a serious game for cultural heritage and in particular for museum environments focusing on the younger generation. The aim of the game is to solve a treasure hunt scenario by collecting medieval objects that used to be located in and around the Priory Undercroft. Located in the heart of Coventry, UK, the Priory Undercrofts are the remains of Coventry's original Benedictine monastery, dissolved by Henry VIII. Initial user testing demonstrated the potentials of serious games for education in museum environments. Game has been created using innovative aspects as far as 3D content reconstruction is concerned as well as educational scenarios. Thus, the main contribution of this paper lies in the direction of accelerating 3D reconstruction by combining computer vision tools and on the educational dimension of the developed game. In addition, artificial intelligence tools are adopted for conducting the plot of the game.

Keywords: serious games, computer graphics, 3D modelling, artificial intelligence, cultural heritage.

1 Introduction

It is well known that we live in a three-dimensional (3D) world and we perceive most of the events/activities/actions via exploiting depth information. Among all

D. Plemenos and G. Miaoulis (Eds.): Intelligent Comp. Graphics 2011, SCI 374, pp. 97–115.
springerlink.com   © Springer-Verlag Berlin Heidelberg 2012

human senses, the visual, and especially the 3D one, is probably the most important in precisely comprehending our world. This abstractive and hierarchical way of thinking of humans is due to "the modularity of mind", as is supported in the pioneering work of Dr. Fodor [1]. According to this theory, our brain follows a hierarchical way of thinking; starting from the simplest thought and ending at the most advanced one. Specifically, this theory implies that if, for example, we want to describe a red rose in a bunch of flowers, the simplest "concept" coming to our mind is . . . an image! To be more specific, it is a 3D image, which clearly consists of all the information we want to describe and which is derived from the experience of seeing a red rose. In other words, a 3D object contains all the ground truth of a physical object, while its 2D views provide only a subset of the 3D object, an abstraction of the real world [2].

3D models have nowadays become ubiquitous for applications such as computer games, robotics, machine vision, computer-aided design, cultural heritage, architectural design and planning [3]-[8]. We live in an era where the acquisition of 3D data is ubiquitous, continuous, and massive. These data come from multiple sources including high-resolution, geo-corrected imagery from aerial photography and satellites. Now researchers are developing mobile geo-located systems-some containing calibrated laser range finders and cameras-that can collect ground-level detail at unprecedented resolution. In the near future, individuals or robots will be equipped with stereoscopic cameras and other sensors as well as the computational power to pervasively collect and organize geo-located data [9].

3D world has generally stimulated different types of applications. Some are related with 3D reconstruction, either precise or approximate and some with computer games technologies. Precise 3D reconstruction (that should resemble as much as possible the real 3D objects) usually demands high degree of humans' interaction that makes 3D computer vision tools be of high cost (manual effort) [10]. Thus it demands huge computational load. For this reason, despite the recent research efforts, 3D computer vision remains one of the most arduous problems in the computer engineering society, especially for real-world applications in which we cannot impose constraints on lighting conditions, object occlusions, and possible dynamic modifications of the environment. This is, for instance, the case of the Future Internet infrastructure which is now moving towards a 3D media cyberspace, 3D content-based visual retrieval and indexing applications, and processes for automatic understanding events and actions in 3D video streams as close as possible to humans' perception [11].

In particular, (i) Future Internet is moving to a *3D media internet*. As the Internet has revolutionized the access to multimedia content and enabled collaborative user-generated content, it is now evolving towards a 3D infrastructure. 3D processing will enable mass distribution and caching of 3D content and enhanced user quality of experience with optimized impact on the performance of the underlying processing and networking platforms. (ii) Despite the current photogrammetric tools for 3D reconstruction, which are tools of high computational cost, 3D media internet needs new "easy and affordable components" for representing the 3D content, algorithms applicable to low-cost devices (of relatively low capabilities) and/or manageable by non expert users. (iii) In the same framework, the recent advances in geographic information systems and in multimedia technologies have made exigent the need for reconstructing 3D data for an extremely vast volume of data (e.g., entire towns). Today, such a 3D digitalization is practically forbidden due to the high computational cost and effort required.

On the other hand, computer games with complex virtual worlds for entertainment have been widespread today, especially amongst the younger generation. These advances have been verified by both the recent achievements in software and hardware technologies. These led to a rise in the quality of real-time computer graphics and increased realism and immersion in computer games.

One particular interesting application of 3D content modelling refers to cultural heritage reconstructions. It is the preservation of historical sites and monuments from erosion, vandalism, and other long-lived artifacts that usually cause damages and need for repair. It's important to keep an accurate record of these sites' current conditions by using 3D model building technology, so preservationists can track changes and/or predict structural problems [10]. The main problem in reconstructing an accurate 3D model is that this process is tedious involving a lot of manual effort and thus making the problem for 3D reconstructing of huge areas (like entire cities) a real-life demanding problem.

Similarly, although the widespread use of gaming for leisure purposes has been well documented, the use of games to support cultural heritage purposes, such as historical teaching and learning, or for enhancing museum visits, has been less considered [12]. As a result, it is important to introduce the concept of serious games to support cultural heritage purposes. The potential use of serious games technologies on cultural heritage education has been recently addressed in the literature and particularly in papers [13][14][15].

A reason for that is the popularity of video games among the younger generation ranging between 10 to 30 years old. Thus, they would be an ideal means for educational purposes. The term 'serious games' describe exactly this concept. The use of a pedagogical game that can be use to teach people under an informal learning scenario. Typical examples are game engines and online virtual environments that have been used to design and implement games for non-leisure purposes, e.g. in military and health training [16][17].

In this paper, we present a new framework for serious games applications in cultural heritage. The developed algorithms allows for real-time interactive visualization and simulation of realistic virtual heritage scenarios, such as 3D reconstructions. Examples include ancient sites and monuments, using off-the-self-components. The serious game is finally evaluated using a small sample of students. Before describing the goals and contribution of this paper, we first cite the most representative works reported in the literature in this theme presenting state-of-the-art papers.

2 Previous Works

Both, entertainment games technologies and serious games technologies share common state-of-art concepts. As the work of [16] mentions, the only difference between serious games techniques and entertainment game structures is the application of games technologies to a non-entertainment domain. However, the difference can be generalized as contributions in the areas of visual expressions, communications and collaboration mechanisms should be properly and promptly examined.

Recent advances on gaming technology are tremendous. The contemporary graphics can achieve in real-time (or at least in a just-in-time framework) near optimal photorealism. This leads to a dramatic increase of virtual games worlds populated with rich multimedia content that considerable improves the quality of experience for the users. And what is common in serious games developers is that it is important to strengthen entertainment, fun and pedagogy [18]. In other words, there is a need for the game developers and instructional designers to work together to develop engaging and motivating serious games for the future.

Moreover, 3D reconstruction for historical sites is an approach that has been studied previously [19][20]. Several funded projects have been developed towards this direction. However, these systems still remain within the academic community without being released to enterprises and/or to wide public for a commercial exploitation.

One of the largest 3D reconstruction projects is the one dealing with the reborn of Ancient Rome. The main aims of the project are to produce a high resolution version of Rome at 320 AD (Figure 1), a lower resolution model for creating a 'mashup' application with 'Google Earth' (http://earth.google.com/rome/), and finally the collaborative mode of the model for use within virtual environments and aimed primarily at education [21]. Towards the same field, the ancient Pompeii (a Roman city, which was destroyed and completely buried in the first recorded eruption of the volcano Mount Vesuvius in 79 AD) was reconstructed [20]. The main goal of this project was to simulate a crowd of virtual Romans exhibiting realistic behaviours in a reconstructed district of Pompeii. Similarly, the Parthenon project reconstructs the Minerva's Temple in the Athenian Acropolis creating a virtual version of the Parthenon and its separated sculptural elements.

Fig. 1 A snapshot of the Rome Reborn project[1]

---

[1] http://www.romereborn.virginia.edu/gallery-current.php

Other types of applications are dealing with the creation of virtual museums using computer games technologies. A recent survey paper that examines all the technologies and tools used in museums was recently published [24]. Here we present several examples of this type of cultural heritage serious game, including some virtual museums that can be visited in real-world museums. A characteristics example includes the game of [25] depicts a hypothetical Virtual Egyptian Temple (no real-world equivalent) embodying all of the key features of a typical New Kingdom period Egyptian temple in a manner that an untrained audience can understand. The game provides enough pedagogical questions that ameliorates the quality of experience for the player and simultaneously increase his/her educational background. In the same framework, the Foundation of the Hellenic World has produced a number of games related with the Ancient Olympic Games [26].

The technology needed to produce all these historical games are based on a game engine which provides the core technology for the creation and control of the virtual world. A game engine is an open, extendable software system on which

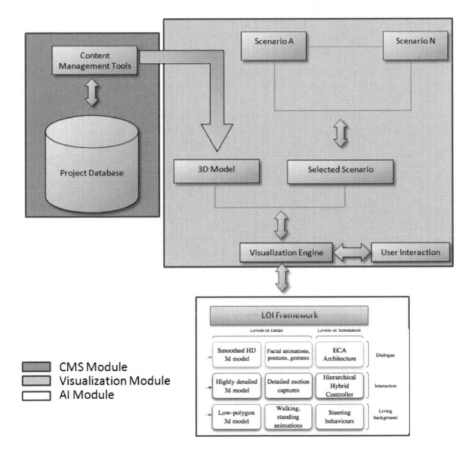

Fig. 2 System Architecture

a computer game or a similar application can be built. It provides the generic infrastructure for game creation [16], i.e. I/O (input/output) and resource/asset management facilities. The possible components of game engines include, but are not limited to: rendering engine, audio engine, physics engine, animation engine.

3D rendering is another important aspect of a serious game, similar to the entertainment games; it requires a lot of graphical features and effects. The state-of-the-art in this subject area is broad and, at times, it can be difficult to specify exactly where the 'cutting edge' of the development of special effects lies. A number of the techniques that are currently in use were originally developed for offline applications and have only recently become adopted for use in real-time simulations through improvements in efficiency or hardware.

3 Game Description

Built on the site of the Benedictine Priory of St Mary's the Centre tells the story of Coventry's first Cathedral, founded in the 11th century by Lady Godiva and Earl Leofric. The centre tells the story of using archaeological finds discovered during the excavation of the site. Located in the heart of Coventry, UK, the Priory Undercrofts are the remains of Coventry's original Benedictine monastery, dissolved by Henry VIII. The Priory Visitor Centre, designed by architects MacCormac, Jamieson and Prichard, brings together objects, images and the latest information about St Mary's Benedictine Cathedral and Priory. Inside the Visitor Centre hangs a mobile of coloured glass by John Reyntiens. It was made using some of the same techniques as the makers of medieval stained glass windows and incorporates the shapes of angel eyes and wings as inspired by the wall painting from the Chapter House. The Priory Undercrofts include remains of the original vaulting, windows and a fireplace.

The motivation is to raise the interest of the younger generation in the museum, as well as cultural heritage in general. The aim of the serious game is to solve a treasure hunt scenario by collecting medieval objects that used to be located in and around the Priory Undercroft. Each time a new object is found, the player is prompted to answer a question related to the history of the site. A typical user-interaction might take the form of: "What did St. George slay? – Hint: It is a mythical creature. – Answer: The Dragon", meaning that the user then has to find the Dragon.

The system architecture, which is presented in Figure 2, is composed by three modules including:

- The Visualization Module
- The Content Management Module
- The Artificial Intelligent Module

The Priory Undercroft Visualization module is based on the Quest3D visualization engine. Quest3D is a very flexible authoring environment for real-time 3D applications. The edit-while-executing and graphical nature of Quest3D makes it one of the most intuitive tools to work with. Quest3D is used by developers, educational institutions and VR companies [27].

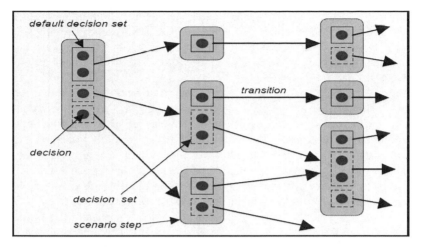

Fig. 3 Decision Tree [28]

The content management system (CMS) is implemented within the authoring tool of QUEST3D and allows the educator to create various learning materials. The CMS is also responsible for the organization of tasks that the user will complete during a particular scenario. Tasks are organized according to a pre-planned

Table 1 Different Puzzles which the player has to solve

Puzzle	Typical Puzzle Question
1	Find the statue of Saint George
2	What did Saint George Slay? Hint: It is a mythical Creature
3	What does a dragon breath
4	Who might have warmed by the fire
5	Monks had had important jobs that took time and precision. What did they create
6	To create perfect illuminated texts the monks had to write in straight lines. They did not use rulers. Can you find the tool they used?
7	What other form of decoration might you find where the monks gathered
8	To show who created a piece of artwork or craft the artist might do what?

decision tree (Figure 3). Each scenario step may contain one or more decisions sets. Each decision set is connected to the next scenario step by a transition that occurs only when all decisions of a given decision set has been taken. In some cases, a decision set may contain a single decision. In this case, the decision leads to the next scenario step.

A decision tree strikes the balance between "user flexibility" and "writer flexibility" [29]. User flexibility, refers to the learner's range of possible actions, in order to make a real-time environment as realistic as possible, this range of action should be maximised. Writer flexibility refers to the developers control over the scenario, ensuring that intended learning outcomes are presented during the scenario. The learner is guided through the scenario with a series of puzzles which they have to solve in order to progress to the next puzzle and finish the games.

Table 1 gives an example of the puzzles that the learner has to solve to progress to the next one. Learner login information is also stored within the database, which will form the basis of a learner tracking system. This tracking system will record decisions and actions made by the user throughout a simulator session for later interrogation by the educator.

The exhibition side is based on the Bergeron [30] principles for a good game interface. Following the guidelines of Bergeron, the exhibition side of the Priory Undercroft was created with the user in mind. The Priory Undercroft graphical interface is illustrated on Figure 4.

Fig. 4 Priory Undercroft Game in Operation

The beginning of the games asks users to select the type of the resolution for their screens. It also asks them whether the game will be played at a window or full screen. It is clear that a user can alter from one mode to the other during the play of the game. Beginning screen is shown in Figure 5. It contains the statistics

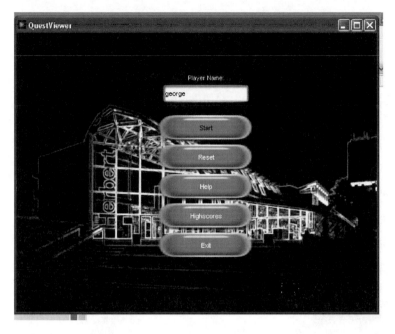

Fig. 5 The first scene of the developed serious game

Fig. 6 Another particular scene of the undercroft reconstructed

for the users play the game and the highest score. It also contains instructions for the game.

Figure 6 shows another scene of the cultural site which we have reconstructed. As mentioned before, the reconstructed was made with a full respect to the archeological evidences Thus, the site is not a fantastic one but a accurate reconstruction of the real objects located within it. In addition, the rules of the serious games were designed to fulfill genuine archaeological evidences so as to provide historical and cultural knowledge to the users via the game.

Figure 7 shows the reconstructed statue of Saint George of this venue. The position of the statue is randomly selected during the game so as to force the users to explore the whole site and thus to get aware of the exact archaeological evidences within this venue.

The educational paradigms adopted are illustrated in Figures 8 and 9. In particular, Figure 8 asks the user to find the St. George statues, while after the user localize the statute the games asks him/her to find what St. George slay?

Fig. 7 Saint George's Statute in the site

The Artificial Intelligence module is based on the 3 Level of Interaction Framework (LoI) which was developed in collaboration between the Serious Games Institute and Toulouse University [31]. The LoI framework simplifies the interaction between the player and the non player characters (NPC). Graphically, the LoI can be represented as auras of increasing complexity centered on the player's avatar (see Figure 10).

Fig. 8 The first educational paradigm

Fig. 9 The second educational paradigm

LoI is based on a simple social space metric [31] and is divided to three levels. The first level aims to populate the characters with authentic crowd in order to increase the immersion of the player. Characters located in closer surrounding of the player belong to the interaction level. Finally, a character inside the dialogue level interacts with the player in a natural way, ultimately using speech recognition and synthesis. All the NPC by default belong to the background level, but as the player moves on the environment and they happen to get closer or away from the player and thus enter or exit the interaction or dialogue levels.

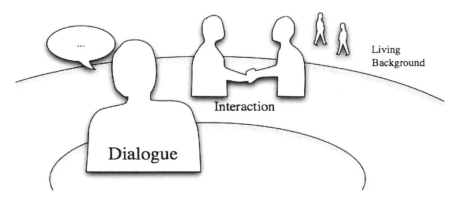

Fig. 10 Level of Interaction Framework

4 3D Reconstruction

Many cultural heritage applications require 3D reconstruction of real-world objects and scenes. This is also the case reconstructing a historical site as in this paper. The recent advances in 3D digitalization and modelling and the recent technological evolutions in laser-scanning, 3D modelling software, image-based modelling techniques, computer power, and virtual reality has permitted the detailed and accurate 3D reconstruction of such places. Many approaches are currently available, mainly exploiting CAD tools and/or traditional photogrammetry with control points and a human operator. However, this approach is time-consuming and can be costly and impractical for large-scale sites. Modelling methods based on laser-scanned data and more automated image-based techniques have recently become available [32].

In our approach, we efficiently combine the traditional laser-scanning approaches with computer vision methods. The latter method is based on the finding of relations among similar geometric patterns in an image. That is, we find the correspondences of characteristic (salient) points between two images that share a part of common content so as to accelerate the time in the reconstruction process.

To acquire data describing an entire structure such as historical site requires taking multiple range scans from different locations that we must register together correctly. Although we can register the point clouds manually, this process is a time consuming and error prone [10]. That is, manually visualization of vast amount of points will surely result in erroneous situations.

Initially, for each image a set of salient points are detected by applying on the visual content scale invariant transformations. Characteristic examples are the Scale-Invariant Feature Transforms (SIFT) which select for any object in an image. Interesting points on the object can be extracted to provide a "feature description" of the object [33]. This description, extracted from a training image, can then be used to identify the object when attempting to locate the object in a test image containing many other objects. The most important issue in SIFT is the fact that the extracted features are invariant under scale, illumination and/or noise alterations.

We also extract the Histograms of Oriented Gradients (HOGs). This technique counts occurrences of gradient orientation in localized portions of an image. This method is similar to that of edge orientation histograms, scale-invariant feature transform descriptors, and shape contexts, but differs in that it on a dense grid of uniformly spaced cells and uses overlapping local contrast normalization for improved accuracy [33].

Let us then denote as $W, V \in R^2$ two sets of two-dimensional (2D) points of the aforementioned features detecting onto two images that share some common parts. Assuming without loss of generality that the two sets include the same number of points and assuming that there exist a known mapping, say A that relates one by one each feature point of set W with each point in set V, then, in statistical theory we can prove that we can estimate the optimal geometric alignment between the two sets.

$$G(W, V; A) = \max_{\mathbf{B}} \sum_{\mathbf{w} \in W} A(\mathbf{B} \cdot \mathbf{w}, A(\mathbf{w})) \tag{1}$$

where $\mathbf{w} \in W$ is a 2D point in set W expressed in homogeneous coordinates. Similarly, B is a transformation matrix. We express the content in homogeneous coordinates since in this space we can get as a matrices product all the affine transformations (scaling, rotation, skewing and translation) including vertical and/or horizontal translation. On the contrary, we use of conventional Cartesian coordinates violates the product property for the simple translation transformation. The main problem of (1) is that it never appears as such in our case, due to unknown feature correspondence and outliers [34].

This problem can be overcome by calculating the correspondences among the two sets $W, V \in R^2$, as we do in finding the disparity field map. In particular, if we denote as $\mathbf{v} \in V$ a 2D point in set V and we recall that $\mathbf{w} \in W$ is a 2D point in set W both expressed in homogeneous coordinates. Then, the correspondences are found by estimating the points in V that best matches with the points in W. In other words,

$$G(W, V; A) = \max_{c} \sum_{\mathbf{v} \in V} \sum_{\mathbf{w} \in W} c_{v, w} d(v, w) \tag{2}$$

In (2) $d(\mathbf{v}, \mathbf{w})$ is an arbitrary similarity measure. This is a very important problem because it can work well in practice if the features are discriminative enough.

It can be proved in [36] that equation (2) can be expressed as an inner product of two independent histograms the contain information about the elements of the two sets. In case that the histograms are normalized this coincides with the cosine similarity measure [37].

To improve the correspondences among the two point sets, we developed in this paper an extension of the RANdom SAmple Consensus (RANSAC) algorithm [38] which yields fast performance relying on the approach of [39]. The work initially refines normals of the point clouds, even in the presence of noise. Then, it computes scores on local planes for each point. Selecting the best local planes and applying a region growing approach we can estimate the correspondences among the different clouds.

5 Procedural Modeling

The game is enhanced with procedural environmental information. This type of information (i.e. trees, street geometry, benches etc) will be implemented to enhance the immersion of the visualisation.

For the terrain itself, heightmaps will be generated using the diamond-square algorithm to provide surface detail. By choosing a recursive algorithm, the level of detail will be adjusted as necessary, which will provide an advantage when dealing with different methods that required different levels of processing power. Instead of using this linear deformation of the neighbouring points, a two-dimensional Lorentz distribution will be used [40]. By adjusting the width of the Lorentzian shape, a means for controlling the smoothness of the terrain can be achieved. Another distribution that can be used as an alternative solution is the Gaussian shape.

- Vegetation: Trees and plants are of significant importance. There are different methods for the procedural creation of vegetation, many of which are based on fractal or simpler rule-based techniques. In this task, a component-based modelling approach [41] will be implemented which provides an intuitive way for controlling plant modelling. The decision of what type of plant needs to be placed into the virtual world usually will depend on a number of factors, including the elevation and slope of the terrain, as well as topographic features that dictate the probability of a specific plant's occurrence [42].
- Buildings and roads: The placement of artificial structures in the virtual world can reach great levels of complexity if the buildings form part of an urban environment [43]. These more complex settlements will be created in a series of steps [44]: (a) first a suitable road network will be generated, effectively providing street maps that partition the terrain and to constrain the placement of buildings (b) this will be then used to direct the division of the terrain into lots which may be partitioned further to generate building footprints, and (c) which will then be used as input for the generation of the buildings themselves.

Another issue is the enrichment of the model with augmented reality scenarios. The augmented reality interface will make 'use' the real environment to superimpose audio-visual and other relevant information to the users. However, the main difference from the previous task is that the content will be provided by the industrial partner, meaning that it won't be generated automatically. The accurate registration of real and virtual information and the effective multimedia augmentation in real time performance are the main challenges of this task:

- Augmented Reality Interface: An interactive and user-friendly AR interface will be developed to achieve maximum presentation using audio-visual information (3D models, spatial sound, 2D images, videos and metadata). As far as the user-machine interaction is concerned simple but effective forms of interaction will be developed but will allow for different modes of operation according to the user needs as well as the case-studies.
- Augmented Reality Tracking: With the reconstructed camera motion and the intrinsic parameters, we can set up the projection matrices. The projection matrix will be composed of the focal length and the principal point, and the modelview matrix will be computed by the camera center position ($C=-R^{-1}T$), orientation (r_3^T: z-axis), and the up-vector (r_2^T: y-axis). When the coordinate systems are matched with our computation and the rendering step, we also have to pay attention to the left-handed and right-handed coordinate systems. In addition, the lens distortion will be considered in this step.
- Augmented Reality Rendering: Initially, the virtual information can be rendered with any computer graphics API (i.e. OpenGL, DirectX, etc). To achieve a high-level of user immersion computer graphics algorithms such as realistic lighting and shading [45], soft and hard shadow generation and occlusions [46] will be considered.
- Human Factors: Extensive user studies are performed to evaluate the cognitive workload as well as presence in the augmented reality environment.
-

Certain "visual gaps" may be derived by the aforementioned methodologies either because of missing information, or due to reconstruction artefacts. These challenges are addressed using novel prediction methodologies able to compensate for the "missing 3D visual" parts exploiting both media, contextual and photogrammetric intelligence. To this end, a set of prediction procedures are developed derived from the input of the previous levels of intelligence. The enhancement process aims to optimize the 3D models generated by the 3D reconstruction stage by replacing the material properties (colouring information and textures) of the 3D objects or re-positioning the geometric topology by taking into account the models derived from the procedural modelling and optimization techniques which smooth and tessellate the 3D data.

6 Game Evaluation

To acquire feedback on the finished core of the application, a self-contained executable file was supplied to a small number of Coventry University students based on the hallway usability testing methodology. The intention of these tests was primarily to gather information on the playability and enjoyability of the game, but also to discover potential technical problems. All of the end-users had some experience with games, and the vast majority described themselves as 'gamers'. A few of those involved also had experience with games programming, or had some knowledge of the architecture behind creating a game. For all users, the aim of the game was presented and it was explained that the players should not expect a complete game, but rather a prototype.

Five students from the Faculty of Engineering and Computing, Coventry University were asked to participate in the test group. The average time of the tests was approximately 30 minutes. Instead of asking the University students specific questions, they were asked to talk through what they were doing and how they felt as they played the game. Overall, recorded feedback was very encouraging and all users agreed that the serious game has a lot of potential for cultural heritage applications. They also mentioned that they prefer the idea of 'playing' and 'learning' at the same time.

On the other hand, a number of important issues were pointed out. In particular, one student had some minor issues with the controls, especially the combination of the mouse and keyboard to navigate inside the environment. It is worth mentioning that after playing the serious game for long enough, the player adjusted to the issue without any further problems. Another user commented that he would play such a game, on the condition that further additions were made to the game play. On the positive side, all users agreed the educational aspect of the game is obvious and helps them to understand and learn something about the history of Priory Undercrofts.

7 Conclusions and Future Work

This paper presented the architecture of a serious 3D game for museum environments focusing on the younger generation. The aim of the game is to solve a treasure hunt scenario by collecting medieval objects that used to be located in and around the Priory Undercroft. Initial user testing demonstrated the potentials of serious games for education in museum environments.

In the future we will add more intelligence to the game, based on intelligent avatars. An online version will be also developed so that it can be accessed remotely. Finally, the game will be installed inside the Herbert Art Gallery & Museum so that it can be evaluated by the museum visitors.

References

[1] Fodor, J.A.: The modularity of mind. Cambridge. Cambridge Bradford Books. MIT Press, Cambridge
[2] Daras, P., Axenopoulos, A.: A 3D Shape Retrieval Framework Supporting Multimodal Queries. International Journal of Computer Vision (2009), doi:10.1007/s11263-009-0277-2

[3] Real-time 3D models, http://www.3drt.com/

[4] Jayanti, S., Kalyanaraman, K., Iyer, N., Ramani, K.: Developing an engineering shape benchmark for CAD models. Computer-Aided Design 38(9), 939–953

[5] Doulamis, N.D., Bardis, G., Dragonas, J., Miaoulis, G., Plemenos, D.: Collaborative Evaluation Using Multiple Clusters in a Declarative Design Environment. In: Artificial Intelligence Techniques for Computer Graphics, pp. 141–157 (2008)

[6] Doulamis, N.D., Bardis, G., Dragonas, J., Miaoulis, G.: Optimal Récursive Designers' Profile Estimation in Collaborative Declarative Environment. In: IEEE International Conferences on Tools of Artificial Intelligence, vol. 2, pp. 424–427 (2007)

[7] Douskos, V., Grammatikopoulos, L., Kalisperakis, I., Karras, G., Petsa, E.: FAUCCAL: An Open Source Toolbox for Fully Automatic Camera Calibration. In: XXII CIPA Symposium on Digital Documentation, Interpretation & Presentation of Cultural Heritage, Kyoto, Japan, October 11-15 (2009)

[8] Douskos, V., Kalisperakis, I., Karras, G., Petsa, E.: Fully automatic camera calibration using regular planar patterns. International Archives of Photogrammetry, Remote Sesnisng and the Spatial Information Sciences 37(5), 21–26 (2007)

[9] Ribarsky, B., Rushmeier, H.: 3D Reconstruction and Visualization. IEEE Computer Graphics and Applications, 20–21 (November-December 2003)

[10] Allen, P.K., Troccoli, A., Smith, B., Murray, S., Stamos, I., Leordeanu, M.: New Methods for Digital Modeling of Historic Sites. IEEE Computer Graphics and Applications, 32–41 (November-December 2003)

[11] http://Cordis.europa.eu//workpackage

[12] Anderson, E.F., McLoughlin, L., Liarokapis, F., Peters, C., Petridis, P., de Freitas, S.: Serious Games in Cultural Heritage. In: The 10th International Symposium on Virtual Reality, Archaeology and Cultural Heritage VAST (2009)

[13] Apperley, T.H.: Virtual unaustralia: Videogames and australias colonial history. In: Proc. of Cultural Studies Association of Australasias Annual, UNAUSTRALIA (2006)

[14] Francis, R.: Revolution: Learning about history through situated role play in a virtual environment. In: Proceedings of the American Educational Research Association Conference (2006)

[15] Jacobson, J., Holden, L.: The virtual egyptian temple. In: ED-MEDIA: Proccedings of the World Conference on Educational Media, Hypermedia & Telecommunications (2005)

[16] Zyda, M.: From visual simulation to virtual reality to games. IEEE Computer 38, 25–32 (2005)

[17] Macedonia, M.: Games Soldiers Play. IEEE Spectrum 39(3), 32–37 (2002)

[18] de Freitas, S., Oliver, M.: How can exploratory learning with games and simulations within the curriculum be most effectively evaluated? Computers and Education 46, 249–264 (2006)

[19] Arnold, D., Day, A., Glauert, J., Haegler, S., Jennings, V., Kevelham, B., Laycock, R., Magnenat, N., Thalamnn, N., Maim, J., Maudu, D., Papafiannakis, G., Thalmann, D., Yersin, B., Rodriguez-Echavarria, K.: Tools for populating cultural heritage environments with interactive virtual humans. In: Open Digital Cultural Heritage Systems, EPOCH Final Event Rome (2008)

[20] Maim, J., Haegler, S., Yersin, B., Mueller, P., Thalmann, D., van Gool, L.: Populating ancient Pompeii, with crowds of virtual Romans. In: VAST 2007: The 8th International Symposium on Virtual Reality, Archaeology and Intelligent Cultural Heritage, pp. 109–116 (2007)

[21] Frischer, B.: The Rome reborn project. How technology is helping us to study history. OpEd. University of Virginia (November 10, 2008)

[22] Debevec, P.: Making "The Parthenon". In: 6th International Symposium on Virtual Reality, Archaeology, and Cultural Heritage (2005)

[23] Jones, G., Christal, M.: The future of virtual museums. In: On-line, Immersive, 3D Environments. Created Realities Group (2002)

[24] Sylaiou, S., Liarokapis, F., Kotsakis, K., Patias, K.: Virtual museums, a survey on methods and tools. Journal of Cultural Heritage (2009)

[25] Jacobson, J., Holden, L.: The virtual Egyptian temple. In: Proccedings of the World Conference on Educational Media, Hypermedia & Telecommunications, ED-MEDIA (2005)

[26] Gaitatzes, A., Christopoulos, D., Papaioannou, G.: The Ancient Olympic Games: Being Part of the Experience. In: The 5th International Symposium on Virtual Reality, Archaeology and Cultural Heritage (VAST), pp. 19–28 (2004)

[27] Quest3D (November 06, 2009), http://www.quest3d.com

[28] Ponder, M., Herbelin, B., Molet, T., Schertenlieb, S., Ulicny, B., Papagiannakis, G., Magnenat-Thalmann, N., Thalmann, D.: Immersive VR decision training: telling interactive stories featuring advanced virtual human simulation technologies. In: EGVE 2003: Proceedings of the Workshop on Virtual Environments, Zurich, Switzerland. ACM, New York (2003)

[29] Magerko, B., Laird, J.: Towards Building an Interactive, Scenario-based Training Simulator. In: 10th Computer Generated Forces and Behavior Representation Conference, Orlando, FL (2002)

[30] Bergeron, B.P.: Developing Serious Games. Game Development Series. Charles River Media, Inc., Massachucetts (2006)

[31] Panzoli, D., Peters, C., Dunwell, I., Sanchez, S., Petridis, P., Protopsaltis, A., Scesa, V., de Freitas, S.: A Level of Interaction Framework for Exploratory Learning with Characters in Virtual Environments. In: Plemenos, D., Miaoulis, G. (eds.) Intelligent Computer Graphics 2010. Studies in Computational Intelligence, vol. 321, pp. 123–143. Springer, Heidelberg (2010)

[32] Sabry, F., El-Hakim, Angelo Beraldin, J., Picard, M., Godin, G.: Large-Scale Heritage Sites with Integrated Techniques. In: IEEE Computer Graphics and Applications, pp. 21–29 (May-June 2004)

[33] David, L.: Object recognition from local scale-invariant features. In: Proceedings of the International Conference on Computer Vision (1999)

[34] Dalal, N., Triggs, B.: Histograms of oriented gradients for human detection. In: IEEE Computer Vision and Pattern Recognition Conference, CVPR (2005)

[35] Dryden, L., Mardia, K.V.: Statistical Shape Analysis. Wiley, Blackwell (1998)

[36] Jegou, H., Douze, M., Schmid, C.: Improving bag-of-features for large scale image search. International Journal of Computer Vision, 1–21 (2010)

[37] Zhu, S., Wu, J., Xia, G.: TOP-K cosine similarity interesting pairs search. In: 2010 Seventh International Conference on Fuzzy Systems and Knowledge Discovery (FSKD), Beijing, China (2010)

[38] David Forsyth, A., Ponce, J.: Computer Vision, a modern approach. Prentice Hall, Englewood Cliffs (2003) ISBN 0-13-085198-1

[39] Deschaud, J.-E.: A Fast and Accurate Plane Detection Algorithm for Large Noisy Point Clouds Using Filtered Normals and Voxel Growing. In: 5th International Symposium on 3D Data Processing, Visualization and Transmission, Espace Saint-Martin, Paris, France, May 17-20 (2010)

[40] Hecht, E.: Optics, 2nd edn., p. 603. Addison Wesley, Reading (1987)
[41] Lintermann, B., Deussen, O.: Interactive Modeling of Plants. In: IEEE Computer Graphics and Applications, vol. 19, pp. 56–65 (1999)
[42] Wells, W.D.: Generating Enhanced Natural Environments and Terrain for Interactive Combat Simulations. Doctoral Dissertation, Naval Postgraduate School, Monterey, CA (2005)
[43] Greuter, S., Parker, J., Stewart, N., Leach, G.: Real-time Procedural Generation of 'Pseudo Infinite' Cities. In: GRAPHITE 2003, ACM SIGGRAPH, pp. 87–94 (2003)
[44] Parish, Y.I.H., Müller, P.: Procedural Modeling of Citie. In: Proc. of ACM SIGGRAPH 2001, pp. 301–308. ACM Press, New York (2001)
[45] Debattista, K., Dubla, P., et al.: Instant Caching for Interactive Global Illumination. Computer Graphics Forum 28(8), 2216–2228 (2009)
[46] Avery, B., Thomas, B.H., Piekarski, W.: User Evaluation of See-Through Vision for Mobile Outdoor Augmented Reality. In: 7th Int'l Symposium on Mixed and Augmented Reality, Cambridge, UK, pp. 69–72 (2008)

Intuitive Method for Pedestrians in Virtual Environments

Jérémy Boes, Cédric Sanza, and Stéphane Sanchez

IRIT, Vortex team
University of Toulouse, 118 route de Narbonne, 31062 Toulouse, France
boes@irit.fr, sanza@irit.fr, sanchez@irit.fr

Abstract. Recent works about pedestrian simulation can actually be sorted in two categories. The first ones focusing on large crowd simulation aim to solve performance and scalability issues at the expense of behavioral realism of each simulated individual. The second ones aim at individual behavioral realism but the computational cost is too expensive to simulate crowds. In this paper, we propose an alternate approach combining a light reactive behavior with cognitive strategies issued from real life videos. This approach aims at the real time simulation of small crowds of pedestrians (one to two hundred individuals) but with concerns for visual realism regarding heterogeneous behaviors, trajectories and positioning on sidewalks.

Keywords: crowd, pedestrians, heterogeneity, small groups, and anticipation.

1 Introduction

At first sight, a real crowd seems to be chaotic and unpredictable. Nevertheless, the local interactions between pedestrians generate auto-organization and emergent structures [4]. For example, parallel lines are formed when many people walk in two opposite directions in a corridor. Moreover, there is a relation between density and average speed of a crowd [15], giving concrete data for computer animations. However, the problem of simulation of a virtual crowd remains quite complex, even if there are many approaches to pedestrian simulation.

Ennis et. al have led a study to find out which criteria are determinant to make a realistic simulation [3]. Most of these criteria are already taken in account by existing simulations, like obstacle avoidance and walking in appropriate areas, but one is not: walking in small groups. Despite it is not mentioned by Ennis, having pedestrians with heterogeneous appearances and behaviors is an obvious key to realism, but not often present in existing models.

In this paper, we present an intuitive approach based on real life observations of pedestrians. The proposed method combines a reactive algorithm of collision avoidance and behavioral strategies. Our goal is to improve visual realism by simulating heterogeneous behaviors and by maintaining small groups of pedestrians.

D. Plemenos and G. Miaoulis (Eds.): Intelligent Comp. Graphics 2011, SCI 374, pp. 117–137.

The second section presents previous works about crowd simulation. The third section presents the model of pedestrian. The third section shows the results of our approach. The last section concludes this paper and gives further works.

2 Related Work

Discrete crowds (also called agent-based simulations) focus on individuals. Local behavioral rules are given to each agent and a realistic global behavior is expected to emerge. The Hidac model [10] uses a combination of psychological and geometrical rules with a social and physical forces model to simulate high-density crowds in normal or panic situations. The behaviors are computed at two levels: the high level behavior (for navigation, learning, communication and decision making) and the low-level motion (for perception, motion and locomotion). A huge number of behaviors are simulated like stopping, queuing, pushing, propagating panic and falling (figure 1a). In [11], Shao simulates a virtual train station. The characters can see all the mobile objects and a limited number of closest mobile objects. They have reactive, motivational and cognitive routines that include low and high level actions like making a purchase or taking a seat. The actions are triggered according to a set of current goals and internal physiological, psychological, or social needs (figure 1b).

Fig 1 A crowd in a complex building using Hidac [10] (left). The waiting room of the Pennsylvania station [11] (right).

Since each agent makes its own decision, discrete crowds allow a great diversity among pedestrians and provide very realistic behaviors, but the computational cost is expensive, which limits the size of the crowd it can handle. Another drawback of discrete crowds is that the local rules are difficult to create. For example, several rules are needed to achieve simple tasks like obstacle avoidance. Since agents have to perceive the world they populate, discrete crowds are dependent of the type of the environment (indoor or outdoor) and of the way it is constructed.

Continuous crowds have a global point of view. Crowd motion is computed with a potential field used by every pedestrian [6]. In [12], Treuille's works are based on three hypothesis: people have a goal, people try to go as fast as possible, people try to avoid areas of discomfort. Thus, the characters move by trying to minimize these three parameters (length of the path, time and discomfort). He uses potential fields to update people's positions from a combination of different grids (density, goals, boundaries…). The model is used in different simulations like evacuation in urban environment (figure 2a).

Recently, aggregate dynamics have combined discrete and continuous models to reach a large number of pedestrians and to handle very dense crowds [8]. The algorithm consists in computing the preferred velocity for each agent and interpolating it with the continuum velocity of the local flow. Finally, he performs collision resolution. The system can manage very dense crowds, until 100 000 characters at 2 frames per second (figure 2b).

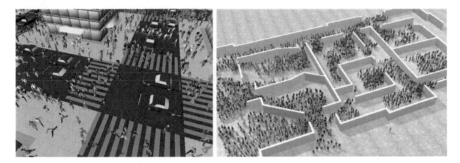

Fig. 2 Evacuation of a building by [12] (left).and [8] (right)

Continuous and aggregate crowds can deal with large dense crowds but are less realistic than agent-based simulations, especially when our eye is caught by one particular character in the simulation. These three approaches are the most popular, but not the only ones existing. The crowd patches method puts together patches of precomputed trajectories [14], it allows to populate infinite worlds

Fig. 3 Animation of crowd with patches [14] (left). Animation of crowd by examples [7] (right).

but virtual humans are not autonomous and the simulation lacks of interactivity (figure 3a).

The crowd by example method constructs a database of situations from the tracking of videos of real crowds (figure 3b). Virtual pedestrians search the database to copy the appropriate trajectory [7]. This method shows very realistic behaviors, but for a small number of pedestrians.

Tables 1 and 2 recapitulate main characteristics and differences amongst the previously described related works.

Table 1 Interactivity means possibilities of manual control or/and on the fly editing of the simulation

	Max number of pedestrians	Real Time Simulation	Interactivity
Shao & Terzopoulos	1200 (without 3D rendering)	Yes (without 3D rendering)	Average
HiDAC	600	Yes	Good
Continuous crowd	10000	No	Low
Aggregate crowds	100000	0 to 10k agents	Low
Crowd patches	3000	Almost	None
Crowds by example	40	No	None

Table 2 Heterogeneity indicates both heterogeneity in characteristics of pedestrians and behaviors. Singularities indicate if the method allows behavioral singularities such as unexpected stops or a pedestrian that do not respect "the rules". Realism means plausibility of simulated crowds

	Heterogeneity	Singularities	Realism	Groups
Shao & Terzopoulos	Average	No	Good	No
HiDAC	Average	Yes	Good	No
Continuous crowd	None	No	Low	No
Aggregate crowds	None	No	Low	No
Crowd patches	Average	No	Low	No
Crowds by example	Low	Yes	Good	No

According to this table we will try to maintain real-time simulations while handling heterogeneous interactive pedestrians walking in small groups. Although scalability is still one of our concerns, the maximum number of simulated pedestrians is not our main priority in this paper.

3 Approach

We focused on three main goals. First, simplicity and genericity: we wanted our method to be easily implemented in any environment. We also wanted it to be able to allow heterogeneous behaviors and small groups of people. To help us to reach a great degree of realism we shot videos of pedestrians walking the downtown streets of Toulouse, France. We extracted precious data from these videos. They are described later. The next section explains how we performed the classic tasks of collision avoidance and retention in walkable areas. The last sections present how we introduced heterogeneity and small groups.

3.1 The Environment

The virtual town project (figure 4) consists of a virtual model of an urban environment in which evolve virtual cars and virtual pedestrians. This environment has been seen in [1] focusing on automatic learning and handling of crosswalks.

Virtual pedestrians are animated using motion captures. Motion realism is not our main concern but behavioral realism and heterogeneity amongst agents are. So each virtual pedestrian is imbued with its own internal parameters such as maximum velocity, weight, height, and perception capacities.

About perception, the pedestrians are able to perceive vehicles, other pedestrians and objects (traffic lights for example) that are in front of them, in their field of view. The field of view is an angular sector which depth and width are set according to the capacities of the pedestrian.

Bullet Physics (http://bulletphysics.org) handles collision detection and their effects on the pedestrians and the objects. The use of physics engine avoids inter-penetration in case of collision of two objects.

Fig. 4 The virtual town project

3.2 Collision Avoidance and Walkable Areas

3.2.1 Obstacle Avoidance

We observed on the videos that pedestrians seem to follow a free space created by those who precede them: they favor *free directions*. We represent these *free directions* in a simple table, called the *direction table*. Each agent has its own table and each cell corresponds to a direction it can take. At any time of the simulation, the value of the cell is the distance the agent can walk, following the corresponding direction, without running into an obstacle (figure 5). The table is initialized with a value that is the maximal distance for which obstacles are taken in account. When an obstacle is perceived, the distance to this obstacle is inserted in the appropriate cell only if it is inferior to the current value of the cell.

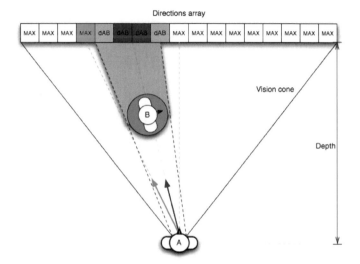

Fig. 5 Agent A perceives agent B on the left. Some cells of the first half of the table are filled with the distance between A and B, the others are empty, they contain the maximal distance. Right arrow represents the aimed direction of the agent A, the left one is the computed direction to avoid agent B.

Each agent computes its *desired direction*. It is the direction that it needs to adopt in order to reach its target. Once its *direction table* is updated according to its perceptions, an agent checks the table if its *desired direction* is free. If not, the agent will have to look for the *closest free direction*. It is the nearest cell containing the maximal distance. The final direction that the pedestrian takes is a weighted average of his *desired direction* and the *closest free direction,* with a greater weight for the latter.

If two cells can pretend to be the closest free direction, the cell with the greater index is chosen. This simulates the natural tendency of people to avoid an obstacle by the right rather than the left when the two solutions are equivalent. In order to obtain smoother trajectories, an agent can adjust its direction even if its *desired direction* is free. This happens when an adjacent cell of the one corresponding to the desired direction contains a small distance. This means that an obstacle is near the trajectory, the agent will then shift its orientation from a cell on the other side in order not to get too close of the obstacle.

The number of cells depends on the angle pitch between each cell. If the pitch is too small, agents don't modify their trajectory strongly enough, if it is too high agents shake and have unnatural trajectories. A pitch of five degrees proved to be the best compromise.

3.2.2 Anticipation

Most of the collisions are easily avoided with this technique (especially with static obstacles), but some still occasionally occur with moving objects. To prevent such collisions, agents don't only perceive size and position of other objects. They also perceive speed and orientation. Therefore they are able to extrapolate the trajectory of other agents. The anticipated position (and not the current position) of perceived obstacles is used to update the *direction table* (figure 6). The amount of time n over which the agent anticipates depends on several criteria: its speed (more he is fast, less he anticipates over a long time), the distance to the other agent (almost no anticipation for very close obstacles) and the angle between the two trajectories (maximum anticipation for perpendicular trajectories, almost no anticipation for parallel trajectories).

Static and moving obstacles are avoided thanks to the same technique, using the same direction table. This allows our method to be easily implemented on any

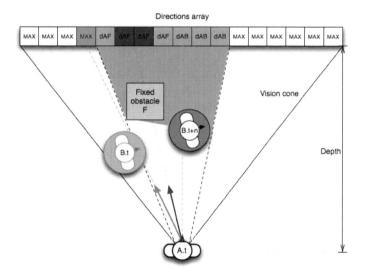

Fig. 6 Agent A extrapolates agent B position. A will avoid B by the left. Without anticipation, A would have turn right and a collision would have occurred for B is moving in this way. The top-left arrow shows the closest free direction cell, the top-right arrow is the desired direction cell.

environment: the only requirement is the perception of distance, position, size, speed and orientation, which is basic. Moreover this technique sticks to reality: if an obstacle stands in our way, we adjust our trajectory just enough to avoid it.

In crowdy environment, the method favors the choice of the less crowdy space in front of the agent (figure 7).

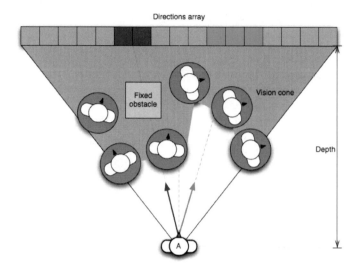

Fig. 7 Example of direction array in crowdy environment. The left arrow indicates the desired direction. The right arrow is the computed one.

At each simulation step, the computation of the direction array is performed once for each agent. Performing only one pass ensures a performance gain but can generate some collisions that are easily handled by the physics engine.

3.2.3 Walkable Areas

To ensure that agents stay on the pedestrian network (sidewalks and crosswalks), we tagged borders with *border cells* (figure 8).

They are perceived by agents and treated as obstacles by the direction table. Agents tend to avoid borders, and stay in safe zones. Border cells are not physical obstacles, if an agent is pushed through a border (it happens when sidewalks are crowded), he will cross it and walk on the road. The direction table allows agents to slightly adjust their trajectory but not to make brutal changes, therefore if a pedestrian walks quickly perpendicularly to a border (it happens if his target is on the road), he will cross it. Of course the treatment of border cells is deactivated for pedestrians who walk on the road.

Fig. 8 Border cells (red) prevent pedestrians to massively walk outside the appropriate areas

3.2.4 Slowing Down and Stopping

Each agent computes an *obstruction rate*, depending on the filling of the *direction table* (number of non-empty cells and average distance). A rate equal to zero means an empty table (no obstacle). Agents slow down if the rate becomes too high, but it will never cause them to stop. The number of non-empty cells has to be taken in account in order to avoid very close but small objects.

$$\tau = \sum_{i=0}^{N} (d_{\max} - d_i) \frac{N^P}{N}$$

Where N is the total number of cells, N^P the number of non-empty cells and d_{max} the maximal distance that it was initialized with.

An agent also slows down when someone walks too close in front of him, approximately at same speed and with the same orientation. Thus, they maintain a personal free space (figure 9).

Agents are able to perceive traffic lights. If it is red for pedestrians, agents willing to cross the street will stop when they arrive at the border of the sidewalk or when they get too close to someone else waiting for the light to turn green.

Fig. 9 Agents waiting at a crosswalk

3.3 Heterogeneity

In real life, crowds are very heterogeneous, both in terms of behavior and of appearance. This diversity is difficult to simulate but is a key to realism. We focused more on the behavior than on the visual aspect. From our observations, we identified three movement strategies: slow strategy, classical strategy and fast strategy.

Slow strategy: People walking slowly are either older persons or people going for a stroll. As they are about 50% slower than classical pedestrians (6 km/h), they do not care about distant obstacles. They only give attention to what is close to them. Their direction table is initialized with a small maximal distance.

Classical strategy: The majority of pedestrians follow this strategy. Classical pedestrians present an average behavior: they stay in appropriate areas, they slow down when too many people are in front of them, but they overtake if someone is too slow.

Fast strategy: Pedestrians that are rushing try to always walk at their maximum speed. They move about 50% faster than classical agents. Their obstruction threshold is higher. Therefore, they slow down less often than classical pedestrians. They are reckless: they don't give attention to border cells so they easily walk on the road if it allows them to overtake a pedestrian or to take a shorter path (figure 10).

Fig. 10 A fast pedestrian overtakes slower agents and walks recklessly on the road

The repartition of these strategies is important in order to get a realistic simulation: a majority of pedestrians must follow the classical strategy. A crowd composed of 80% classical, 10% slow and 10% fast pedestrians gave good results. These strategies bring heterogeneity and singular behaviors to the simulation. Fast agents do not respect the usual rules, like some people in real life.

3.4 Small Groups

In real life, we observe that more than half the people walk in small groups of two to six pedestrians. We counted on our videos 726 pedestrians, 43% of them walk alone, while 32% walk in pairs, 18% in groups of three people, 7% in groups of four people and the last 2% in groups of five or six people (figure 11).

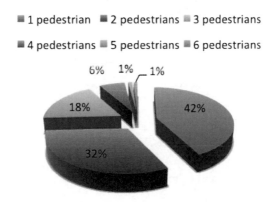

Fig 11 Repartition of pedestrians computed from video observations

In our model, groups consist of a leader and followers. The leader decides of the speed and the direction of the group, followers copy their behavior on him (figure 12).

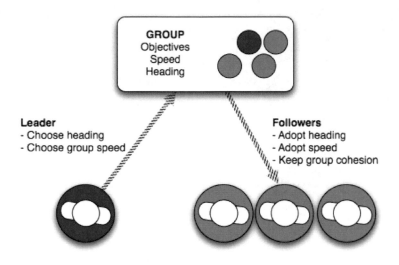

Fig. 12 Interaction between group structure and agents

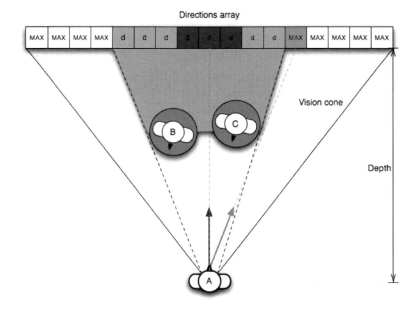

Fig. 13 Agents B and C are part of a group, agent A sees them as a single obstacle, he will not walk between them

They all share the same objectives. How to combine obstacle avoidance and group cohesion is an open question. For now, a group does not perform obstacle avoidance with moving objects. Alone pedestrians perceive groups as a single obstacle. They try to not cut through it (figure 13 and 14).

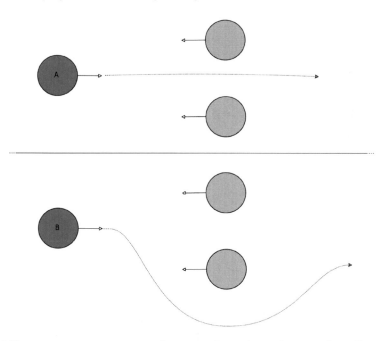

Fig. 14 Up, grey agents are not part of a group. Agent A cuts between them. Down, grey agents are part of a group. Agent B avoids them.

Each agent stores in its memory a list of other agents he knows. If an agent who walks alone (or is the leader of a group) meets one of them during simulation, they both will stop, stand a few seconds face to face and finally form a group. The leader of the new group is chosen arbitrarily. Fast pedestrians never stop when they meet a friend and do not form groups.

4 Results

4.1 Experimental Protocol

We have run several experiences. In each one of them we put several pedestrians in our virtual town. Each agent has an individual navigation path along the sidewalks. They must stay as much as possible on the sidewalks and cross the streets when they are allowed to.

In order to evaluate the pertinence of our choices, we ran series of tests. Each test was based on the same principle: two simulations were launched initialized the same way, but one of them had a deactivated feature or different parameters.

In the following screenshots, the textures of pedestrians indicate their behavior and characteristics as described in table 3.

Table 3 Characteristics and strategies according to textures (when activated in simulations)

Texture	Maximum speed	Strategy
White shirt, white mane	Slow	Slow
Black shirt, gray pants	Medium	Classical
Military outfit	Fast	Classical
Red jacket, black pants	Fast	Fast
Black jacket, blue cap	Very Fast	Fast

4.2 Collision Avoidance

The first test concerned collision avoidance. We ran two simulations with fifty pedestrians initialized the same way (same positions, same strategies repartition, no small group), but agents of one of them did not perform collision avoidance (figure 15).

Fig. 15 Left, collision avoidance is deactivated. Right, collision avoidance is activated.

It was visually obvious that having no obstacle avoidance ruins the realism. We also followed five agents on each simulation, during one minute, and counted how many times a collision occurred with one of them. Twenty-five collisions occurred when avoidance is off. Only two occurred when it is activated (table 4).

Table 4 Collisions per pedestrian, one-minute simulation, 40 agents

Pedestrian	A	B	C	D	E	Total
With avoidance	0	0	1	0	1	2
Without avoidance	4	4	6	6	5	25

A usual observed (and realistic) feature of crowd simulation, and more specifically, of opposite flows of walkers, is the formation of lanes [2,5,16] due to collision avoidance. The pedestrian lanes consist of pedestrians that share the same intended direction and approximately the same velocity [13].

Those lanes are also presents in our simulation (figure 16).

Fig. 16 Lanes formation in opposite flows

4.3 Walkable Areas

We tested the visual impact of retaining pedestrians in appropriate areas with two simulations. Both were initialized with the same pedestrians at the same positions, one of the simulations had *border cells* but the other had not. The result is shown by figure 17: it is obvious that realism is enhanced when agents walk where they are supposed.

4.4 Small Groups

The impact of the presence of small groups was evaluated by the comparison between a simulation where every pedestrian is alone and another where some of them walk together, in small groups. The simulation with small groups seemed more natural (figure 18).

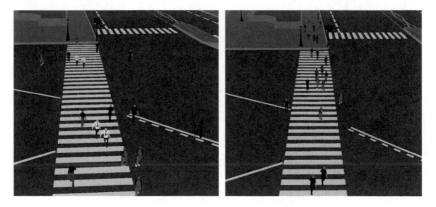

Fig. 17 Left, with no border cell, agents don't stay on sidewalks and crosswalks. Right, agents mainly stay on safe areas. It is obviously more realistic.

Fig. 18 Left, a simulation where everybody walks alone. Right, some agents walk in pairs or in groups of three persons.

4.5 Heterogeneous Pedestrians and Strategies

To measure the effect of strategies heterogeneity, we ran a simulation where every pedestrian follows the classical strategy and another where the 80-10-10 repartition was respected. The first one gave the impression of a "clone army" since everyone was moving at the same pace (figure 19). To enhance visualization, agents have been colored in accordance with their initial position. Left, a series of screen shots shows the evolution of a simulation where every agent follows the classical strategy. The crowd seems "frozen". Right, 80% of the agents follow the classical strategy, 10% the slow strategy and 10% the fast strategy. There is more "mixing" between pedestrians, as fast agents overtake slow agents.

Fig. 19 Left, only classical strategy. Right, heterogeneous strategies

Heterogeneity in agents and strategies enriches the simulation, bringing singular behavior (very slow pedestrians, agents walking on the road, etc.).

4.6 Plausibility

According to our results, pedestrians smoothly avoid static and moving obstacles. They stay in appropriate areas. They have different behaviors and are able to form small groups.

Moreover, the different strategies allow singular perturbations like pedestrians crossing the street when and where they should not or walking out of sidewalks (figure 20).

Fig. 20 Left, a fast pedestrian walks out of sidewalk. Right, agents walks out of crosswalk

Thus, the main Ennis criteria to bring great realism to a simulation seem fulfilled.

To give a general idea of final results, figure 21 compares a real scene with a simulated one.

Fig. 21 Left, a real life photograph from downtown Toulouse. Right, a simulated scene in similar conditions

4.7   Implementation and Performance Considerations

The *direction table* technique is light and intuitive. It can easily be implemented in any model allowing obstacle detection, with no need for any sophisticated environment. Besides, each pedestrian being driven by an individual behavioral engine (to execute its strategy), one can change its behavior or manual control it at any time in the simulation.

On a computer with an Intel Core 2 Duo 2,4GHz and 2 GB RAM our simulation can handle up to fifty agents without any lag at 30 frames per second (including 3D rendering). With 75 agents the framerate is down to 20 fps, we manage to keep an interactive rate (10 fps) with 200 agents.

4.8 Synthesis

The tables 5 and 6 indicate our contribution relatively to discussed related work in section 2.

Table 5 Interactivity means possibilities of manual control or/and on the fly editing of the simulation

	Max number of pedestrians	Real Time Simulation	Interactivity
Shao & Terzopoulos	1200 (without 3D rendering)	Yes (without 3D rendering)	Average
HiDAC	600	Yes	Good
Continuous crowd	10000	No	Low
Aggregate crowds	100000	0 to 10k agents	Low
Crowd patches	3000	Almost	None
Crowds by example	40	No	None
Our work	200	Yes	Good

Table 6 Heterogeneity indicates both heterogeneity in characteristics of pedestrians and behaviors. Singularities indicate if the method allows behavioral singularities such as unexpected stops or a pedestrian that do not respect "the rules". Realism means plausibility of simulated crowds

	Heterogeneity	Singularities	Realism	Groups
Shao & Terzopoulos	Average	No	Good	No
HiDAC	Average	Yes	Good	No
Continuous crowd	None	No	Low	No
Aggregate crowds	None	No	Low	No
Crowd patches	Average	No	Low	No
Crowds by example	Low	Yes	Good	No
Our work	Good	Yes	Good	Yes

5 Conclusion

In this paper, we have presented our works based on an intuitive approach for pedestrians. The main characteristics of our system are to manage heterogeneous behaviors and cohesion of small groups. In existing methods, these two principles are few used although they substantially increase the realism of the simulations.

Nevertheless, the system can be improved in several ways. The small groups could manage children with own behavior (running, returning back to the parents). The strategies could be continuous, from slow to fast, in order to get more heterogeneous behaviors. The optimization of the direction table is possible by sharing information from close agents moving in the same direction.

Future works focus on the number of simulated characters by using level of details at two levels: graphical and especially behavioral. In our group, we develop a system enabling to manage the LOD of the interactions with the objects [9,17]. This system could help us to simulate more characters and several complex actions.

References

[1] Abdul Karim, A., Sanza, C.: Learning by implicit imitation in virtual worlds. In: CASA 2010, Workshop on Crowd Simulation, Saint-Malo (2010)
[2] Daamen, W., Hoogendoorn, S.P.: Experimental research of pedestrian walking behavior. In: Transportation Research Board Annual Meeting, pp. 1–16 (2003)
[3] Ennis, C., Gerdelan, A., O'Sullivan, C.: Plausible Methods For Populating Virtual Scenes. In: CASA (2010)
[4] Helbing, D., Molnar, P., Farkas, I.J., Bolay, K.: Self-organizing Pedestrian Movement. Environment and Planning B: Planning and Design (2001)
[5] Helbing, D., Buzna, L., Johansson, A., Werner, T.: Self-organized pedestrian crowd dynamics: Experiments, simulations, and design solutions. Transportation Science 39(1), 1–24 (2005)
[6] Hughes, R.L.: The Flow of Human Crowds. Annual Review of Fluid Mechanics 35 (2003)
[7] Lerner, A., Chrysanthou, Y., Lischinsky, D.: Crowds by Example. In: Eurographics (2007)
[8] Narain, R., Golas, A., Curtis, S., Lin, M.C.: Aggregate Dynamics for Dense Crowd Simulation. In: Siggraph Asia (2009)
[9] Panzoli, D., Peters, C., Dunwell, I., Sanchez, S., Petridis, P., Protopsaltis, A., Scesa, V., de Freitas, S.: A Level of Interaction Framework for Exploratory Learning with Characters in Virtual Environments. In: Plemenos, D., Miaoulis, G. (eds.) Intelligent Computer Graphics 2010. SCI, vol. 321, pp. 123–143. Springer, Heidelberg (2010)
[10] Pelechano, N., Allbeck, J.M., Badler, N.I.: Controlling Individual Agents in High-Density Crowd Simulation. In: ACM Siggraph/Eurographics Symposium on Computer Animation (2007)
[11] Shao, W., Terzopoulos, D.: Autonomous Pedestrians. ACM Siggraph/Eurographics Symposium on Computer Animation (2005)
[12] Treuille, A., Cooper, S., Popovic, Z.: Continuum Crowds. In: Siggraph (2006)
[13] Usher, J.M., Strawderman, L.: Simulation of Emergent Crowd Behavior Using Microsimulation of Individual Pedestrians. Computers & Industrial Engineering 59, 736–747 (2010)

[14] Yersin, B., Maïm, J., Pettré, J., Thalmann, D.: Crowd Patches: Populating Large-Scale Virtual Environments for Real-Time Applications. In: Proceedings of Symposium on Interactive 3D Graphics and Games (2009)
[15] Weidmann, U.: Transporttechnik der Fußgänger. IVT (90) (1993)
[16] Weng, W.G., Shen, S.F., Yuan, H.Y., Fan, W.C.: A behavior-based model for pedestrian counter flow. Physica A: Statistical and Theoretical Physics 375 2, 668–678 (2007)
[17] Zertal, S., Djedi, N., Sanza, C., Sanchez, S., Duthen, Y.: Exploitation des niveaux de détails dans la simulation du comportement d'humains virtuels. In: 1st International Conference on Information Systems and Technologies (2011)

Pathfinding with Emotion Maps

Anja Johansson and Pierangelo Dell'Acqua

Dept. of Science and Technology, Linköping University
60174 Norrköping, Sweden
{anja.johansson,pierangelo.dellacqua}@liu.se

Abstract. In humans as well as animals, emotions greatly influence decision making on both a conscious and subconscious level. In this paper we use findings within psychology and neuroscience to create a model for modeling and using emotion maps for agent pathfinding. We propose a model that combines these emotion maps depending on the agent's current emotions. We aim for a more varied and more natural agent behavior. Our hope is to create interesting behavior for games or game-like scenarios.

Keywords: artificial intelligence, emotions, emotion maps, pathfinding.

1 Introduction

Our work focuses on intelligent autonomous virtual characters in dynamic virtual worlds. We focus mainly on non-player characters (NPCs) in games or game-like simulations. The aim is to create characters that behave in an interesting and dynamic way, as this will improve the player experience. Within our project an extensive agent architecture has been developed for this purpose.

Emotions affect human decision-making in numerous ways. While this has been studied extensively in psychology, it has to a large extent been neglected in the field of artificial intelligence. Up until the last decade, emotions have been treated as noise or potentially dangerous additions to AI. Recently there has, however, been an upswing in the use of emotions in agent architectures. Previously, we have introduced emotions into our decision making module [11]. The pathfinding has not used emotions, however.

This paper describes our method of using emotions for pathfinding purposes. We introduce the concept of emotion maps. These are created given the agent's emotional state at different locations in the environment. The agent can then use the emotion maps to affect the pathfinding in various ways. The motivation for this method is to create a more interesting agent behavior. NPCs in games such as role-playing games (RPGs) could greatly benefit from such a pathfinding. We also discuss how it would be possible to use these emotions maps when choosing the target position.

D. Plemenos and G. Miaoulis (Eds.): Intelligent Comp. Graphics 2011, SCI 374, pp. 139–155.
springerlink.com   © Springer-Verlag Berlin Heidelberg 2012

The structure of this paper is as follows. First, we will provide background information regarding emotions and their impact on spatial memory. Second, we will describe previous work within emotional pathfinding. Third, we will describe our method. Finally we will show the results and round off with some general conclusions.

2 Background

Originally, studies concerning decision making excluded emotions due to their seemingly irrational nature. The past two decades, however, studies within psychology have brought the importance of emotions to the surface. The work by Damasio [4, 5] in particular changed the view of emotions in the field. Together with Bechara *et al.* [2] Damasio suggests a somatic marker hypothesis. In short, this hypothesis proposes that the decision making process is influenced by marker signals which appear in bioregulatory processes. Emotions and feelings are parts of these processes.

Generally speaking, emotions play a fundamental role in human decision-making ([4, 5, 14, 17, 31]) both as a tie-breaker between numerous choices but also as a quite direct and anticipatory factor. In general, humans make decisions that try to maximize the positive emotions while minimizing the negative [17, 18, 21]. It is therefore worth noticing that not only current emotions affect the decision making, but previous and expected emotions as well.

Studies by Lerner *et al.* [15, 16] show that emotions greatly influence the perceived risk of an action. The studies also show that happy and angry people are more optimistic, while fearful people are more pessimistic. It is interesting to note that emotions with the same valence can affect decision-making in very different ways. Similar results have been found by Raghunathan *et al.* [29].

Humans can associate objects, events and other types of information with emotions on a subconscious as well as a conscious level [14]. Even people with impaired conscious long-term memory can create these types of subconscious emotional memories and these will affect how the person behaves the next time he/she is exposed to a similar situation. LeDoux [14] makes a distinction between the two memories; the implicit, subconscious memory is called an *emotional memory*, while the explicit, fact-like, conscious memory is called a *memory of an emotion*. An emotional memory can trigger a body response (e.g. the triggering of fear when being exposed to a situation that resembles a past bad experience) while a memory of an emotion is merely a declarative memory, a fact. However, the strength of the conscious memory is affected by the emotions felt at the time one acquired it. A memory with stronger emotions attached to it will be forgotten more slowly.

Studies [27] have shown that animals who are placed in a context, in which they previously have been exposed to uncomfortable experiences, will react negatively to that context. This may include the place at which the unpleasant events occurred as well as neutral objects, animals, sounds, etc. present at the time. This sort of emotional conditioning has been studied in humans as well. In 1920, Watson *et al.* [33] conducted a series of experiment on a human baby to determine how emotional responses can be triggered to neutral objects and events by emotional conditioning. In recent years, fear conditioning in particular has been of great interest. For a summary of more recent studies on fear conditioning, see work by Maren [20].

It is known that people who have experienced something traumatic in a certain place often avoid that place afterwards. In cases of post-traumatic stress, this tendency can be pushed to the extreme, where the victims experience intense anxiety, merely by being exposed to a situation similar to the traumatic event [13, 35].

In the field of environmental psychology many studies (for a review of these studies, see the work by Manzo [19]) have been conducted concerning emotional attachment to places. While many of the studies are focused on the notion of "home", some discuss more general place attachment situations [34]. It is also apparent that people can have negative attachment as well as positive attachment to places.

Given this information about the way emotions affect our choices and reactions, it is easy to deduce that emotions affect the routes we choose, subconsciously and consciously, when we move in our environment. Most people will, when choosing a route in their environment, choose a route that has previously been more pleasant to them. For instance, if a person has witnessed an accident in one place, the memories of that accident may cause the person to avoid that place both on a conscious and subconscious level. A less dramatic example is choosing a more scenic route (that has previously made the person feel good) above a boring or stressful one.

3   Previous Work

Surprisingly, emotions have not been widely used for pathfinding purposes. There is to best of the authors' knowledge no commercially available computer game that uses emotional pathfinding. Within the academia, few articles have addressed the subject. In this section we will present work from different fields that attempt to do emotional pathfinding in one form or another.

Christian Nold [25] uses emotion maps for artistic purposes. In the experiments, the test subjects are equipped with a measurement device that measures the level of physiological arousal. The subjects are then asked to walk around in a given neighborhood. The results are mapped onto a real map of the area. The subjects are also asked to clarify the reason behind their emotional response at different locations on the map. The resulting map is used mainly as a visualization of the area, but can be used for tourists to find an interesting route to take. The downside of this method is that the output from the measuring devices cannot be mapped directly to emotions, but rather to physiological arousal. Therefore, part of the process requires manual input.

To the best of our knowledge, in the field of computer science only two groups have used emotions for pathfinding. Donaldson et al. [7] have made an attempt to do emotional pathfinding. They let the currently most important emotion of the agent decide where it wants to go. Unfortunately, they do not actually use emotions much. Rather, they use physiological states, such as hunger and thirst, and let these decide where to go. They modify the A*-algorithm so that emotions affect the heuristic function. It is, however, unclear exactly how they do this.

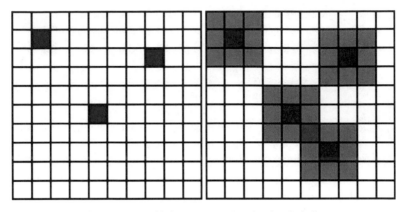

Fig. 1 a) Object positions on the grid, b) An example of a simple influence map

Mocholí *et al.* [24] propose an agent architecture that uses emotions for decision-making and movement within the world. Their system focuses on emotional virtual agents in an augmented reality application. The agents are capable of attaching emotional memories to other entities within the world, or to human bystanders [1]. The emotions affect where the agents choose to go and what they do. The pathfinding is a mixture of graph-based search and a particle system. The result is a set of possible paths. Which path is chosen is decided by the emotional behavioral layer.

4   Emotion Maps

For the construction of emotion maps we are inspired by the work on *influence maps* [23, 32]. Although the concept of influence maps is vaguely defined in general, we choose to define it as any information represented as a 2D environmental map that is used to assist the agent's decisions, or in our case, the pathfinding process. Influence maps have been used mostly in strategy games for high-level decision making purposes but are suitable for other types of decision making scenarios as well [22].

An influence map in our system is represented by a two-dimensional uniform rectangular grid. An example of an influence map can be seen in Figure 1 b). This influence map is generated from the position of objects seen in Figure 1 a). This is a simple influence map, where only the surrounding cells are affected by the nearby object cell.

Our wish is to create a set of emotion maps that show in what areas an agent has experienced certain emotions. We create one emotion map for each emotion that we find relevant for pathfinding. Furthermore, we assume that an area around an *emotional position* (a position at which the agent experiences emotion) is affected. We represent this by using a falloff function with respect to the distance from a point on the map to the emotional position.

Moreover, small emotional values should not affect the map as much as large emotional values. Suppose a place has been visited by the agent several times and each time the agent experiences a small bit of happiness. We do not want these small emotional experiences to add up to a large value on the map. We therefore filter the emotional values using a sigmoid function depicted in Figure 2.

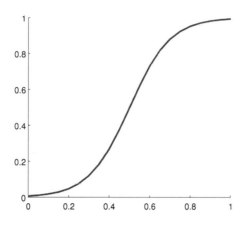

Fig. 2 A sigmoid filter with slope value 0.1 and horizontal offset 0.5

Formally, an emotion map map_e for emotion e is constructed and updated with respect to the agent's current position p in the following way:

- For each position X in the map map_e
  - Calculate the euclidean distance d between X and p.
  - Filter the emotional value e_{val} using a sigmoid:

$$weight = \frac{1}{1 + e^{-(e_{val}-0.5)/b}} \qquad (1)$$

  where b is the slope steepness. e_{val} is the current value of the emotion e.
  - Calculate a strength S as[1]:

$$S = 1 - \frac{1}{1 + e^{-(d/rad-0.5)/s}} \qquad (2)$$

  where rad determines the influential radius each emotional position has on the map and s is the slope steepness.
  - Multiply S by e_{val} and $weight$.
  - Add S to the value at position X.

In simpler terms, this results in a map where every emotional position gets a small radius of influence around it. The emotion map is updated continuously, to allow the new emotional and positional information to be added to it.

---

[1] One may choose other falloff functions if desired. We choose this sigmoid-related function due to its smooth and predictable nature.

In our architecture, emotions are represented as a sum of sigmoid signals (for more information, see work by Johansson *et al.* [9]). However, any emotion representation that uses continuous emotion values between 0 and 1 can be used together with our model.

5   Using Emotion Maps

Let us assume we have calculated emotion maps for the following emotions: *sadness*, *happiness* and *fear*. What does this tell us about the different positions on the maps?

Little research has been done on exactly how specific emotions affect the choice of routes. This is most likely due to the fact that emotions cannot be measured precisely. There are several ways to estimate which emotions a person feels. One can measure physiological states, such as skin conduction and heart rate, but it is not straightforward to map these into what we perceive as emotions [6]. There may be reasons for the state of the body that has nothing to do with the experienced emotions, for instance an increased heart rate because of exercising. One can also measure expressive facial reactions [8, 12] such as frowning or smiling but the results may vary depending on context and social situation [28]. Another popular approach is to let people rate their emotions according to measurement scales [30]. This is also imprecise and highly subjective. While all these measurement techniques have merit when one wishes to get a general idea of the emotional state of a person, they do not produce accurate numerical values of a person's emotions.

Because there are no studies on exactly how emotions affect the choice of routes, we will use an intuitive approach to decide how to combine the different emotion maps.

5.1   Combining Emotion Maps

When combining the emotion maps, we want to be able to weight the maps differently, with both negative and positive weights. We know that emotions can influence decision making and memory with different amounts depending on the emotion. For instance, fear is a more important factor than both happiness and sadness and its weight should therefore be of higher magnitude.

The agent's current emotions should affect how the emotion maps are merged together. It is known that the current emotional state affects the strength of recollected events [3]. E.g. a happy memory is more easily remembered when a person is happy.

The procedure to calculate the emotion maps is straightforward. First, normalize[2] the emotion maps so that the values range between 0 and 1. Second, calculate the combined emotion map *CM* out of a set of emotions maps *E* as follows:

---

[2] While this may result in over-emphasized areas on the map at the beginning of the simulation, it makes further calculation more controllable.

$$SUM = \sum_{i \in E} map_i * w_i * (1 - a + a * emo_i) \qquad (3)$$

where map_i is the normalized emotion map for the emotion i and w_i is the weight for the emotion i. The constant a is used to determine how much the current emotions should affect the merging of the emotion maps. emo_i is the emotional value for the emotion i at the time the combined emotion map is calculated.

$$CM = \frac{SUM - w_{sumMin}}{w_{sumMax} - w_{sumMin}} \qquad (4)$$

where w_{sumMax} is the maximum possible value that CM can take, which is the sum of all the positive weights w_i, with $i \in E$. w_{sumMin} is the minimum possible value, which is the sum of all the negative weights.

The combined emotion map should ideally be recalculated every time it is used, i.e. every time pathfinding is performed so that it uses the latest emotion maps and the agent's current emotional state.

5.2 Choosing a Goal and Finding the Path

Let us assume the agent is hungry and wants to grab something to eat. It knows of four different positions on the map where there are cafés. Which one should it choose? One can use the emotion maps in many ways to aid the agent's decision. In our method, given the combined emotion map in 5.1, we look at the values of the combined map at each of the four positions. The position with the highest value is the one the agent would choose if it does not have to take into consideration other things, such as distance to cafés, time-constraints, what areas it would have to walk through to get there, etc.

Once the goal has been decided, the combined map in Section 5.1 can be used as an influence on the pathfinding. We use a straightforward version of the A*-search algorithm on a hierarchical node structure. For more information on how this model works, see work by Johansson *et al.* [10].

We let the combined map influence the cost needed to move between points on the map. The cost function of the A*-algorithm is defined as:

$$f(n) = g(n) + h(n)$$

where $g(n)$ is the cost of the calculated path from the start node to n and $h(n)$ is a heuristic estimate of the cost of the remaining path to the goal node. $g(n)$ can be obtained by summing up the costs between nodes in the path (from start node to n). We define the cost $f(n_1, n_2)$ between two nodes as

$$f(n_1, n_2) = c + 1 - c * \frac{CM(n_1) + CM(n_2)}{2} \qquad (5)$$

where c is a constant that signifies how much the emotion maps should be taken into account during the pathfinding process. The minimum possible value of $f(n_1, n_2)$ is 1, which is the equivalent of using the distance as a measurement of the cost.

For heuristics, $h(n)$, we use euclidean distance to the goal since it is difficult if not impossible to redefine the heuristics function to include the emotion map somehow. Also, the euclidean distance will always be optimistic with respect to the actual cost, and hence fulfills the optimality requirements of the A*-algorithm. This will, however, result in a lack of efficiency during searches if the cost function is much larger than the heuristics[3].

6 Results

In this section we will demonstrate our method with two examples and show the results.

There exist many theories concerning basic emotions (see summary by Ortony *et al.* [26]). The models differ greatly in terms of number of basic emotions and which emotions that are counted as basic. We calculate the following emotion maps; *sadness*, *happiness* and *fear*. Should one wish to include other emotions or relevant physiological states such as stress, one can do so. The reason we choose these three emotions is that we believe they are more tightly connected to geographical places than many other basic emotions (of those commonly listed), such as e.g. *anger*.

To set up our first experiment, we set up an environment where an agent can walk freely. We fill the environment with items that the agent will react emotionally to (see Figure 3 d))[4]. There are three types of objects; those that will make the agent happy (green), sad (blue) and afraid (red), respectively. The agent is free to explore the environment as it wishes. While in this experiment we use only the visibility of items as triggers of emotions, the algorithm is in no way limited to this. Any emotions felt by the agent, for any reason, will be used to create the emotion map.

6.1 Emotion Maps

The emotion maps are calculated continuously during the simulation. The size of the maps in our simulation is 128 by 128. The emotion maps at the end of the simulation are depicted in Figure 3 a), b) and c). To illustrate them together, a colored version (where happiness is green, sadness is blue and fear is red) is shown in Figure 3 f). The combined influence map calculated from there three emotion maps using the method described in Section 5.1 is depicted in Figure 3 e).

To calculate these maps and the combined map, the following parameters were used. For filtering the emotional values (see Equation 1), we use $b = 0.1$. To calculate the strength (Equation 2), we use $rad = 30$ and $s = 0.2$. The weight values (see Equation 6.5) for *happiness*, *sadness* and *fear* were 1, -2 and -4 respectively.

---

[3] For our purposes, speed has not proved to be a problem since we use a hierarchical node structure that speeds up the search considerably.

[4] Please note that the reason behind the agent's emotions is irrelevant. They are in no way limited to the appearance of objects in the environment. We only do this in the example for simplicity.

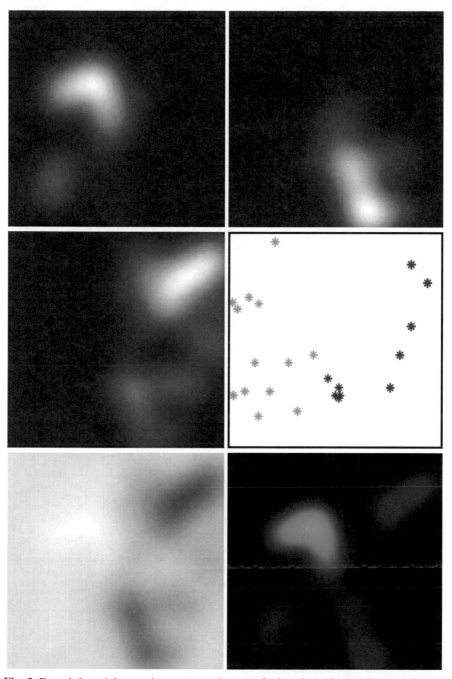

Fig. 3 From left to right, top down: a) emotion map for happiness, b) emotion map for sadness, c) emotion map for fear, d) objects colored according to emotional attachment e) the combined emotion map, f) a colored version where the emotion maps have different colors.

To calculate the emotion maps in Figure 3, we used no emotional impact ($a = 0$ in Equation 6.5).

The maps calculated were 128 by 128 in grid size, encompassing a 200 units by 200 units simulation environment.

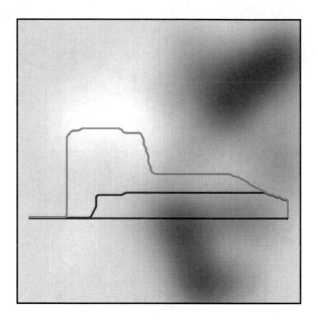

Fig. 4 Resulting paths with different amount of influence from emotion maps. The blue path uses no emotion map influences. The red path uses $c = 3$. The green path uses $c = 4$.

6.2   Paths

In Figure 4 three different paths are shown. These paths have been generated using different amounts of the combined emotion map seen in Figure 3 e). One can see that when the emotion maps are not taken into account, the agent walks straight to its target. Depending on the strength of the parameter c in Equation 5, the agent will walk more or less in areas that is more pleasant to it.

Please note that the seemingly jerky turns of the path is due to the pathfinding using 4-connectivity in the grid.

6.3   Emotional Impact

To illustrate how the emotional impact (a in Equation 6.5) works, a series of combined maps are depicted in Figure 5. From left to right, top down, the figures show the change in the combined map when using emotional influence. We use full emotional impact ($a = 1$) and let the agent's emotions *fear* and *happiness* vary. *sadness* is kept at a constant value of 0.5.

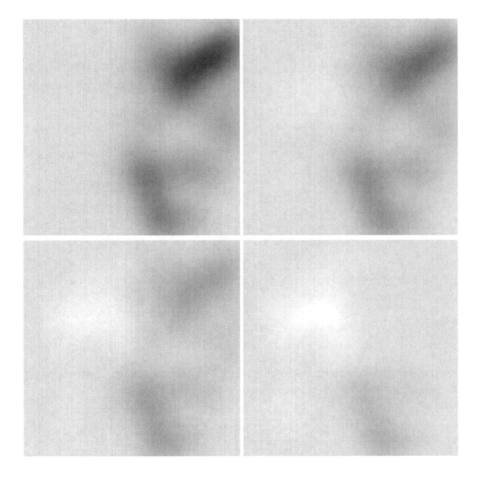

Fig. 5 Emotion maps with full emotional impact with different emotional values. From left to right, top down: a) $fear = 1, happiness = 0$, b) $fear = 0.66, happiness = 0.33$, c) $fear = 0.33, happiness = 0.66$, d) $fear = 0, happiness = 1$. Sadness has a constant value of 0.5.

One can easily see that when the agent is happy, the emotion map for happiness dominates the agent's combined map. When the agent becomes afraid and accordingly also less happy, the emotion map for fear is the map that dominates.

6.4   Choosing Target

We want to test how the emotion maps influence the agent's choice of target. We let the agent choose between four different positions on the map. These positions can be seen as the black dots in Figure 6. The values for the combined emotion map were (left to right, top down) 0.895, 0.312, 0.888 and 0.707. The largest value is the value that signifies the most desirable destination. This means that the agent will choose the top-left target.

Note that this is a simple example and does not take into consideration if the agent is in a hurry, where in the world it is, what other needs it has, etc. These results merely illustrate the use of emotion maps as a part of choosing target position.

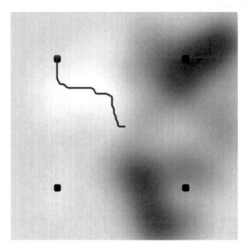

Fig. 6 The agent can choose between four targets. The combined emotion map can be seen in the background. The resulting path is shown in blue.

6.5 Obstacles

To illustrate how our model functions with the presence of obstacles in the scene, we set up another test environment. In this environment, we introduce walls that the agent cannot walk through (see the red areas in Figure 7 e) and f)). We let the agent explore this environment freely. After running the simulation for a period of time, the emotion maps for *happiness*, *fear* and *sadness* can be seen in Figure 7 a), b) and c) respectively. A colored version of the maps can be seen in Figure 7 f). We use the same parameter settings as in the first test example (see Section 6.1). The wall thickness of the obstacles is roughly 5 world units[5].

First, the agent chooses a path without using current emotional values. In such a case, only the weights of the different emotions (w_i in Equation) determine how the emotion maps are merged. This will favor the emotion map for *fear*, because we use a high constant for *fear*. The merged map can be seen in the background of Figure 7 e). The resulting path will go through the corridor in the upper part of the map. When we let the agent's current emotions affect the pathfinding ($a = 1$), a different combined map is created (background of Figure 7 f)). In this case, the

---

[5] The thickness of the obstacles is relevant when one considers the effect of emotions "bleeding" across the walls. We address this issue further in the discussion.

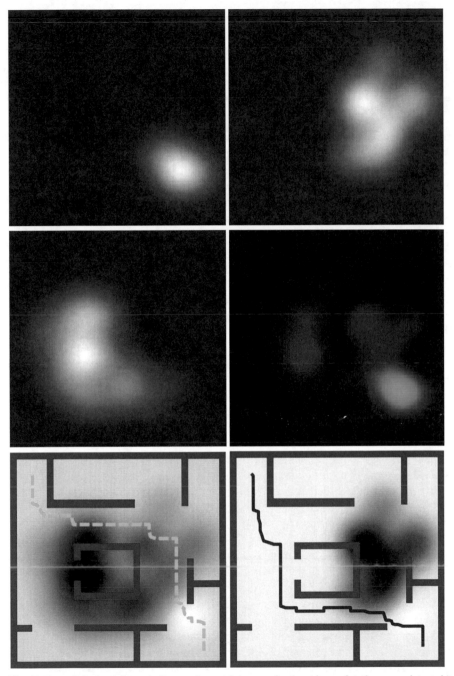

Fig. 7 From left to right, top down: a) emotion map for happiness for the second test, b) emotion map for sadness, c) emotion map for fear, d) a colored version where the emotion maps have different colors, e) the resulting path when no current emotions are taken into account, f) the resulting path when the agent experiences sadness.

agent has a high level of sadness (1.0) and a little bit of happiness (0.3). The agent is unafraid. This in turn will make the agent choose a path that goes through the lower corridor. The behavior mimics how emotions affect memory retrieval as mentioned in Section 5.1.

7 Discussion

In this paper we use findings within psychology and neuroscience to create a model for modeling and using emotion maps for agent pathfinding. We have also proposed a way to dynamically merge the emotion maps depending on the agent's current emotions. Furthermore, we have shown the result of using our model in an environment where the agent is exposed to emotional situations. The benefit of our model is that the behavior of the agent, in terms of pathfinding, is more varied and more natural. This is a potential interesting addition to games, such as RPGs, where single characters are clearly visible and need to be interesting.

As mentioned in Section 3, little research has been done concerning emotional pathfinding in artificial intelligence. In contrast to the work by Donaldson et al. [7], our model uses emotions directly in the cost-function of the pathfinding. While their work focused mainly on physiological states, we focus on emotions. Compared to the work by Mocholí et al. [24] our system uses a quite different pathfinding process. While their pathfinding process generates several possible paths that are either accepted or rejected by an emotional decision making system, our system works in a more dynamic way, automatically choosing an emotionally suitable path from the start. Our model also works in a subconscious manner, letting the agent's emotions affect the agent's decisions without awareness or reasoning about the cause of the emotions.

It is worth noticing that the emotions map are unique to each agent. Should one have several agents in the simulation, the emotion maps generated by each agent will differ, sometimes quite drastically. This reflects the different events the agents have been exposed to, their personalities, how they react to events and what choices they have previously taken when choosing routes through the environment. This individuality gives increases the potential for interesting NPCs.

Setting up the parameters for the calculations is fairly straightforward. The radius of influence, rad in Equation 2 should be set to a reasonable size depending on the relative unit size of the environment (a unit in the simulation vs. a meter in the real world). Parameters that determines the influence of current emotions on the calculations, such as a and c in Equation 4 and 5 should be tuned to receive the desired behavior in different situations and to suit different agent personalities. The parameter c is greatly influenced by the context the agent is in. If the agent is in a hurry to reach a target, it may decide to walk through a negative area to get there on time. However, if the agent is not in a hurry, it may take a more pleasant route.

The computational complexity of calculating an emotion map for an emotion is $O(n)$, where n is the number of grid cells in the emotion map. The reason for this is that we continuously update the emotion map with each new update of the agent (once a second or more often). Because of this, only the current position of the agent accompanied with the emotional state of the agent must be used. The actual processing time is 0.015 seconds on an Intel Pentium Dual Core 2.80GHz processor[6]. Should one wish to recalculate the entire emotion map given memories of old position of the agent, naturally the computational time will be much higher.

When using obstacles, it is worth noticing how the emotions "bleed" over to the other side of the walls. If this is acceptable from a psychological point of view is questionable. On the other hand, it is most likely that equal distances are not treated equally by the mind depending on the context. More studies are needed to determine how humans perceive emotional places in terms of distance and obstacles.

A limitation as well as a benefit of our method to combine the emotion maps is the lack of conscious reason behind the emotions. Since the emotions are mapped onto a two-dimensional grid, it is neither possible or relevant to reason about the source of these emotions. This, on the other hand, makes certain scenarios impossible. For instance, if an agent experiences fear due to a very specific reason and that reason has been removed, the agent cannot adjust its decision making to take this into account. The subconscious emotional memory mentioned in Section 2, however, cannot be reasoned away and thus our method behaves correctly in that respect.

Our model could be expanded further in several ways. Currently, it does not implement forgetting of any kind. Once an agent has experienced an emotion at a certain location, the memory of that emotional experience will never decrease in intensity. A possible extension is introducing forgetting in the model in some way.

It is worth noting that the example of emotionally choosing targets mentioned in Section 6.4 is merely an example to demonstrate the full use of emotion maps. When choosing targets, the emotion maps should be merged with another form of measurement, such as distance to targets, to better simulate real situations. There are many factors to consider. Is the agent in a hurry? Will it spend a relatively long time at the target to make up for having to go through a negative emotional place to get there? We leave the developing of sophisticated methods for merging different measurements when choosing targets as future work.

References

1. Acosta, R., Esteve, J.M., Mocholí, J.A., Jaén, J.: Ecoology: An emotional augmented reality edutaiment application. In: International Conference on Cognition and Exploratory Learning in Digital Age (CELDA), Barcelona, pp. 19–26 (2006)
2. Bechara, A., Damasio, H., Damasio, A.R.: Emotion, decision making and the orbitofrontal cortex. Cerebral Cortex 10(3), 295–307 (2000)
3. Bower, G.H.: Mood and memory. American Psychologist 36(2), 129–148 (1981)
4. Damasio, A.R.: Descartes' Error: Emotion, Reason and the Human Brain. Harper Perennial (1995)
5. Damasio, A.R.: The Feeling of What Happens: Body and Emotion in the Making of Consciousness. Harcourt Brace & Co., New York (1999)

---

[6] Only one of the cores were used for calculating the map, however.

6. Desmet, P.M.: Measuring emotion: Development and application of an instrument to measure emotional responses to products. In: Karat, J., Vanderdonckt, J., Blythe, M., Overbeeke, K., Monk, A., Wright, P. (eds.) Funology. Human Computer Interaction Series, vol. 3, pp. 111–123. Springer, Heidelberg (2005)
7. Donaldson, T., Park, A., Lin, I.-L.: Emotional pathfinding. In: Webb, G.I., Yu, X. (eds.) AI 2004. LNCS (LNAI), vol. 3339, pp. 31–43. Springer, Heidelberg (2004)
8. Ekman, P.: Facial expressions of emotion: New findings, new questions. Psychological Science 3(1), 34–38 (1992)
9. Johansson, A., Dell'Acqua, P.: Realistic virtual characters in treatments for psychological disorders - an extensive agent architecture. In: Proceedings of SIGRAD 2007: Computer Graphics in Healthcare, pp. 46–52. Linköping University Electronic Press (2007)
10. Johansson, A., Dell'Acqua, P.: Knowledge-based probability maps for covert pathfinding. In: Boulic, R., Chrysanthou, Y., Komura, T. (eds.) MIG 2010. LNCS, vol. 6459, pp. 339–350. Springer, Heidelberg (2010)
11. Johansson, A., Dell'Acqua, P.: Affective states in behavior networks. In: Plemenos, D., Miaoulis, G. (eds.) Intelligent Computer Graphics 2009. SCI, vol. 240, pp. 19–39. Springer, Heidelberg (2009)
12. Kaiser, S., Wehrle, T.: Facial Expressions as Indicators of Appraisal Processes, In Appraisal processes in emotion, ch. 16, pp. 285–300. Oxford University Press, Oxford (2001)
13. Kassam-Adams, N., Fein, J.A.: Posttraumatic stress disorder and injury. Clinical Pediatric Emergency Medicine 4(2), 148–155 (2003)
14. LeDoux, J.E.: The Emotional Brain. Simon and Schuster, New York (1996)
15. Lerner, J.S., Keltner, D.: Beyond valence: Toward a model of emotion-specific influences on judgement and choice. Cognition and Emotion 14, 473–493 (2000)
16. Lerner, J.S., Keltner, D.: Fear, anger, and risk. Journal of Personality and Social Psychology 81, 146–159 (2001)
17. Loewenstein, G., Lerner, J.S.: The Role of Affect in Decision Making. Oxford University Press, New York (2003)
18. Loewenstein, G., Schkade, D.: Wouldn't it be nice? predicting future feelings. In: Kahneman, D., Diener, E., Schwarz, N. (eds.) Hedonic Psychology: Scientific Approaches to Enjoyment, Suffering and Well-Being. Russell Sage Foundation, New York (1999)
19. Manzo, L.C.: Beyond house and haven: toward a revisioning of emotional relationships with places. Journal of Environmental Psychology 23, 47–61 (2003)
20. Maren, S.: Neurobiology of pavlovian fear conditioning. Annual Review of Neuroscience 24, 897–931 (2001)
21. Mellers, B., Schwartz, A., Ritov, D.: Emotion-based choice. Journal of Experimental Psychology: General 128(3), 332–345 (1999)
22. Miles, C., Louis, S.J.: Towards the co-evolution of influence map tree based strategy game players. In: 2006 IEEE Symposium on Computational Intelligence and Games (CIG), pp. 75–82 (2006)
23. Millington, I., Funge, J.: Artificial Intelligence for Games, 2nd edn., vol. 6. Morgan Kaufmann, San Francisco (2009)
24. Mocholí, J.A., Esteve, J.M., Jaén, J., Acosta, R., Xech, P.L.: An emotional path finding mechanism for augmented reality applications. In: Harper, R., Rauterberg, M., Combetto, M. (eds.) ICEC 2006. LNCS, vol. 4161, pp. 13–24. Springer, Heidelberg (2006)
25. Nold, C.: Emotional Cartography - Technologies of the Self (2009),
    http://www.emotionalcartography.net
26. Ortony, A., Turner, T.J.: What's basic about basic emotions. Psychological Review 97(3), 315–331 (1990)

27. Phillips, R.G., LeDoux, J.E.: Differential contribution of amygdala and hippocampus to cued and contextual fear conditioning. Behavioral Neuroscience 106(2), 274–285 (1992)
28. Picard, R.W.: Affective computing. M.I.T Media Laboratory Perceptual Computing Section Technical Report No. 321 (2000)
29. Raghunathan, R., Tuan Pham, M.: All negative moods are not equal: Motivational influences of anxiety and sadness on decision making. Organizational Behavior and Human Decision Processes 79(1), 56–77 (1999)
30. Richins, M.L.: Measuring emotions in the consumption experience. The Journal of Consumer Research 24(2), 127–146 (1997)
31. Slovic, P., Finucane, M., Peters, E., MacGregor, D.G.: The affect heuristic. European Journal of Operational Research 177(3), 1333–1352 (2007)
32. Tozour, P.: Game programming gems 2, ch. 3.6. Cengage Learning (2001)
33. Watson, J.B., Rayner, R.: Conditioned emotional reactions. Journal of Experimental Psychology 3(1), 1–14 (1920)
34. Williams, D.R., Patterson, M.E., Roggenbuck, J.W., Watson, A.E.: Beyond the commodity metaphor: Examining emotional and symbolic attachment to place. Leisure Sciences 14, 29–46 (1999)
35. Yehuda, R.: Post-traumatic stress disorder. New England Journal of Medicine 346(2), 108–114 (2002)

Gait Evolution for Humanoid Robot in a Physically Simulated Environment

Nesrine Ouannes[1], NourEddine Djedi[1], Yves Duthen[2], and Hervé Luga[2]

[1] LESIA Laboratory/ Med Khider Biskra University, Departement of computer science, BP 145 RP Biskra, Algeria
{nesrine.ouannes, noureddine.djedi}@lesialab.net
[2] Institut de Recherche en Informatique de Toulouse, Université de Toulouse - CNRS - UMR 5505 Toulouse, France
{Yves.Duthen,Herve.Luga}@univ-tlse1.fr

Abstract. This article describes a bio-inspired system and the associated series of experiments, for the evolution of walking behavior in a simulated humanoid robot. A previous study has demonstrated the potential of this approach for evolving controllers based on simulated humanoid robots with a restricted range of movements. The development of anthropomorphic bipedal locomotion is addressed by means of artificial evolution using a genetic algorithm. The proposed task is investigated using full rigid-body dynamics simulation of a bipedal robot with 15 degrees of freedom. Stable bipedal gait with a velocity of 0.94 m/s is realized. Locomotion controllers are evolved from scratch, for example neither does the evolved controller have any a priori knowledge on how to walk, nor does it have any information about the kinematics structure of the robot. Instead, locomotion control is achieved based on intensive use of sensory information. In this work, the emergence of non-trivial walking behaviors is entirely due to evolution.

1 Introduction and Motivation

Simulation of human motion has applications in several different fields like, entertainment industry, education, science visualization, architecture and medicine. Different research areas, with different aims, are involved in the study of the human motion to understand the fundamental dynamics and its control mechanisms. The human body comprises 206 bones, over 600 muscles and is controlled by a complex nervous system. The human motion is the result of 92 degrees of freedom. Researchers in biomechanics, robotics, and computer science work to understand human natural motion in order to reproduce it artificially.

The aim of humanoid robotic researchers is to obtain robots that can imitate the human behavior to collaborate, in the best way, with humans. Building a complete humanoid robot is a very complex task and researchers usually prefer solving simpler problems as the study of the biped robots, artificial hands, vision, questions about high level control etc.

D. Plemenos and G. Miaoulis (Eds.): Intelligent Comp. Graphics 2011, SCI 374, pp. 157–173.
springerlink.com
© Springer-Verlag Berlin Heidelberg 2012

There are numerous application areas for robots with anthropomorphic shape and motion capabilities. In a world where man is the standard for almost all interactions, such robots have a very large potential acting in environments created for people. They can function in certain areas which are not accessible for wheeled robots, such as stairways or uneven terrain. Furthermore, robots capable of bipedal locomotion have the ability to interact with the environment using the whole body, and climb over large obstacles in their path as opposed to wheeled vehicles. The major drawback of legged robots is the challenge of creating controllers for them. The problem is complicated because of the number of degrees of freedom in each leg and because of changes in the body center of mass and momentum.

In the research literature, and on topics related to bipedal walking, the terms stability, equilibrium, and balance are often used inter-changeably. Throughout this paper, we will use the following notions in order to avoid confusion; the term stability refers to a system which could be analytically treated as stationary (that is the static case), whereas for a non-stationary system (the dynamic case), the terms balance and equilibrium are used.

An obvious problem confronting humanoid robotics is the generation of stable and efficient gaits. Whereas wheeled robots normally are statically balanced and remain upright regardless of the torques applied to the wheels, a bipedal robot must be actively balanced, particularly if it is to execute a human-like, dynamic gait. The success of gait generation methods based on classical control theory, such as the zero-moment point (ZMP) method [23], relies on the calculation of reference trajectories for the robot to follow.

In order to address this problem, alternative, biologically inspired control methods have been proposed, which do not require the specification of reference trajectories. The aim of this paper is to describe one such method, based on recurrent neural networks (RNNs), for control of bipedal robot.

2   Related Work

Various methods were proposed to generate bipedal robot locomotion using evolutionary techniques. Peterson [18] has reported on the development of a method for generating walking behaviors for bipedal robots. An adaptation of evolutionary programming (EP) to the case of nite state machines (FSMs) is used to operate on both the structure and the parameters of the robotic brain. The method has been demonstrated on a simplied ve link robot, constrained to move in the sagittal plane. Two test cases were used; energy optimization and robust balancing. Typically, robotics researchers employ bio-inspired control strategies based on articial neural networks (ANNs) [14, 26] or central pattern generators (CPGs) [5]. Often some kind of evolutionary algorithm (EA) is utilized for the design of the controller [18, 19, 2, 27], and [28].

Of course, the proposed technique would have a greater impact if demonstrated in a 3-dimensional rigid-body simulation instead. There exist, however, some examples in the research literature of synthesizing locomotion for full rigid-body simulated bipeds, by incorporating EAs.

Evolutionary computation, frequently involving the evolution of neural network controllers, has been successfully used to the automation of the process of gait creation [12, 11, 8, 20, 25, 6, 24, 15].

In recent years Jeff Clune [4] demonstrates that HyperNEAT, a new and promising generative encoding for evolving neural networks, can evolve quadruped gaits.

Recently [1] Amin Azarbadegan describes the design of an approach to evolve Simss creatures with morphology and behavior similar to biped animals, his hypothesis is that biases in morphology that encourage limb specialisation, combined with rewards for successful locomotion and carrying at the same time and realistic, physics-based penalties for falling in fitness function, would lead to creatures capable of bipedal locomotion.

He evolved bipedalism by incorporating physical damage and incentives for upright locomotion. The reward for carrying is reflected in the components of the fitness function involving keeping the head up, limiting the number of limbs and making two limbs exempt from damage.

In this paper we propose an approach that uses evolutionary techniques with neural network to evolve a controller for an humanoid robot that is physically simulated.

3   Models and Methods

3.1   Humanoid Robot Model

he humanoid robot model used here was created from the body and joint primitives available in ODE[1] . simulation package. To build the kinematical model, the geometrical data of the robot (link lengths, type and position of joints etc.) are needed.

A physically-based model of legged locomotion describes the nonlinear relationships between the forces and the moments acting at each joint and the feet etc., and the position, velocity and acceleration of each joint angle. In addition to the geometrical data, a dynamics model requires kinematical data as mass, centre of mass and inertia matrix for each link and joint, max/min motor torques and joint velocities which are difficult to obtain and are often an overlooked source of simulation inaccuracy. To simulate interaction with the environment detection and handling of collisions as well as suitable models of foot-ground contacts are required. In the context of simulation of autonomous robots for research purposes often uses the open source Open Dynamics Engine (ODE). ODE can handle collision detection for several geometric primitives.

The biped model used here is a fully tree-dimensional bipedal robot with 15 degrees of freedom as shown in the right most panel of fig.1. The robot model consists of 12 rigid-body parts, and 11 ODE joints. There are 1 fixed joint, 8 hinge joints and 2 Ball/socket joints used to connect the rigid-body parts into an articulated rigid-body structure (hinge, fixed and ball/socket are the internal names of specific joint types in ODE). The rigid-body primitives used are 8 capped cylinders, 3 rectangular boxes and one sphere. The robots structure is defined using multiple chains,

---

[1] Open Dynamic Engine.

Table 1 Body parameters of the humanoid robot

Body part	Geometry	Dimension (m)
Head	Sphere	Radius:0.188
Arm	Caped cylinder	$0.14 \times 0.25 \times 0.44$
Torso	Rectangular box	$0.90 \times 0.25 \times 1.00$
Thigh	Caped cylinder	$0.20 \times 0.25 \times 0.70$
Shank	Caped cylinder	$0.20 \times 0.25 \times 0.70$
Foot	Rectangular box	$0.40 \times 0.50 \times 0.10$

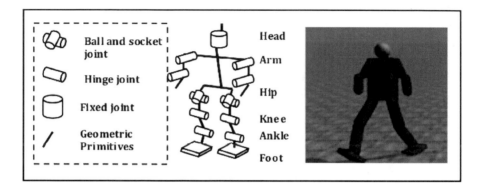

Fig. 1 I3-D biped model used in our simulation

starting from its feet with each link described in terms of the previous links. This composition results in a 15 degree of freedoms (DOFs) bipedal model.

3.2   Locomotion Control Using Recurrent Neural Network

A practical biped needs to be more like a human:

- Capable of learning new gaits when presented with unknown terrain. In this sense, it seems essential to combine force control techniques with more advanced algorithms such as adaptive and learning strategies. Conventional control algorithms for humanoid robots can run into some problems related to mathematical tractability, optimization, limited extendibility and limited bio-logical plausibility. The presented intelligent control techniques has the potential to overcome the aforementioned limitations;
- Furthermore, some researchers have begun considering the use of neural networks for control of humanoid walking [5, 17, 16, 14, 26]. This approach makes possible the learning of new gaits that are not weighted combinations of predened bipedal gaits. Various types of neural networks are used to generate walking behaviors and control design of humanoid robots such as multilayer perceptron, CMAC (Cerebellar Model Arithmetic Controller) networks, recurrent

neural network, RBF (Radial Basis Functions) networks or Hopeld networks, which are trained either by supervised or unsupervised (reinforced) learning methods. The majority of the proposed control algorithms have been veried by simulation, while there were few experimental verications on real biped and humanoid robots. Neural networks have been used, as efficient tools for the synthesis and off-line and on-line adaptation of biped walk. Another important role of connectionist systems in controlling of humanoid robots has been their ability to solve static and dynamic balance during the process of walking and running on terrain with different environmental characteristics.

3.2.1   Recurrent Neural Network (RNN)

An RNN is an artificial neural network [9] consisting of a number of neurons (nodes) with arbitrary connections (including self-coupling of individual neurons) (see figure 2). This network can operate either in discrete time as common in feed-forward networks (i.e. ANNs without feedback connections), or in continuous time. In the later case, using a simple neuron model, the dynamical behavior of the ith node in the network is governed by the equation:

$$\tau_i + \gamma_i = \sigma(\beta_i + \omega_{i\varphi}\gamma_i + \omega_{i\varphi}^I I\varphi)._i = 1, 2, ..., \nu \qquad (1)$$

Where ν is the number of neurons in the network, τ_i are time constants, γ_i is the output (activity) of node φ, $\omega_{i\varphi}$ is the (synaptic) weight connecting node ν to $node_i$, $\omega_{i\varphi}^I$ is the weight connecting in-put node ν to $node_i$, $I\varphi$ is the $\nu^I h$ external input to $node_i$, and β_i is the bias term, which determines the output of the neuron in the absence of inputs. $\sigma()$ is a sigmoid function whose main purpose is to restrict the activity of the neurons to a given range. The hidden, context, and output layers of the RNN all use the same bipolar sigmoid activation function (see equation 2 and fig. 3).

$$\sigma(c) = 2/(1 + e^{-\alpha c}) - 1. \qquad (2)$$

The obvious advantage an RNN has over the traditional feed for-ward network is "memory." The use of feedback connections allows the RNN to have a "memory" of past events. Thus, pattern presentation to the RNN will take into consideration what moment in time the pattern occurs. Biological neural networks process information in a similar fashion to the RNN.

The humanoid robot presented in this paper uses Elman (1990) [7] Recurrent Neural Network for its biological plausibility and powerful memory capabilities. Furthermore, biological neural net-works do not make use of back propagation for learning. Instead we use evolutionary algorithms to evolve locomotive behaviors.

In difficult real-world learning tasks, such as controlling robots, it is intractable to specify correct actions for each situation. In a complex control system, such as that used by the humanoid robot, specifying the correct outputs for each possible input combination and state is practically impossible. In these situations, optimal behavior must be learned by the exploration. In that case the reinforcement of good

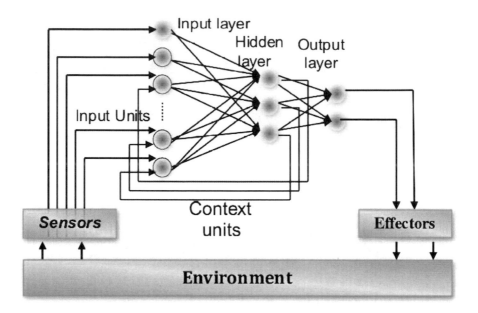

Fig. 2 The sigmoid bipolar function

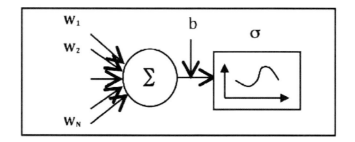

Fig. 3 The sigmoid bipolar function

decisions relies on some feedback from the system itself and the exploitation of learned knowledge from the environment. Genetic algorithms can be used as an optimization process to evolve neural networks that prove to be robust solutions to difficult real-world learning tasks without the need to supply additional information or for an external agent to direct the process.

3.3   The Model of RNN

The humanoid robot uses a set of sensors to collect data from the environment and feeds it to the RNN. Sensors monitor the internal state of the robot, such as joint angles are referred to as proprioceptive sensors. In this setting, the current joint angles of the previous time step of the simulation are used by the evolved controller to compute the next set of motor signals for the robot.

Simulating a biped robot in a realistic environment most likely re-quires feedback loops between the robot's control system and the robot's body, as well as between the control system and the environment.

The set of external sensors constitute the robot's "window" upon the environment. Those sensors can measure values such as the robot acceleration or inclination relative to a fixed coordinate frame, light intensity, external forces applied to the robots body, etc. The arrangement of the sensors is shown in figure 3. In this figure, the humanoid robot has 15 sensors located throughout its body. The contact sensors indicate when the feet make contact with the ground plane, and the angle sensors measure the angle of each hinge joint of the robot body. The Ball and socket joints also contain angular velocity sensors that feed the rate of angular change back to the RNN. And finally, a direction sensor provides the ANN with a virtual compass.

3.4 Evolutionary Optimization of RNNs

In the standard genetic algorithm (GA) [10], which is one example of an EA, the variables of the problem are encoded in a xed-length string. By contrast, the EA used here acts directly on the RNNs.

The topology of our network is made up of 4 layers: an input layer, a hidden layer, an output layer and a context layer. The number of neurons contained in the input and output layer depends on the robots morphology (i.e. the number of sensors and effectors). Each in-terneuron connection within the RNN is assigned a weight.

Real-number encoding is used, i.e., all genes take oating point values in the open interval [0,1], which are then rescaled to the appropriate range.

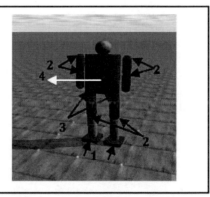

| 1: Touch sensors |
| 2: Angle sensors |
| 3: Angular velocity sensors |
| 4: Direction sensor |

Fig. 4 Sensor arrangements

3.4.1   Genetic Algorithm

Genetic algorithms are a class of search and optimization heuristics inspired by Darwins theory of evolution. They operate on a population of individuals (potential solutions to a problem), updating the population in parallel, over many generations. With its mechanism, the genetic algorithm has proven to be a useful tool with large search spaces (such as, for example, to find the best parameters for locomotion). In genetic algorithms, the Fitness Function is of extreme importance: it is the operator that, during the evolution, evaluates all the individuals (their phenotypes) of the current generation. The concept of "the best individual" depends on the problem. So, in the case of this work, a good phenotype is a set of parameters that causes a character to perform the desired movement. In a GA, the best individuals will be more likely to pass their genes to the individuals in the next generation. The fitness function drives the evolution and its definition is crucial for the algorithms performance: with a poorly designed fitness function the GA might miss good solutions. In this work, the purpose of the genetic algorithm is to optimize the weights of the neural network which controls the humanoid robot. A synergistic relationship exists between the GA and the RNN. The GA optimizes the RNN, and the RNN produces robot behavior that is then scored. This feedback will then drive the GAs population to converge to an optimum. At start-up, the population's chromosomes are initialized at random. The chromosome length is 1130 genes, with 1 gene per RNN weight. The number of connections represents the number of genes in the chromosome; a floating-point number represents each gene. To avoid the problem of premature convergence, a linear fitness function is used in this paper. The GA used here makes use of the standard single-point crossover operator. After the crossover operation, the gene has a probability of being mutated. The mutation operator uses a Gaussian perturbation rather than a random mutation. By perturbing the weights rather than randomly selecting values for the mutated weights, we favored gradual change. Table II presents the static parameters used for the GA.

3.4.2   The Fitness Function

The fitness function determines how an individual is rated in terms of genetic fitness, and indirectly influences the agent behaviors. The fitness function was carefully chosen such that it would tend to award efficient locomotive behaviors and penalize wasted effort. It is also based on the distance travelled by the robot within a certain period of time. A higher fitness score is awarded to humanoid robots that are able to travel larger distances in a given amount of time (10 seconds).

The reward is the Euclidean distance between the robots center of mass at its initial position and when it eventually fell.

4   Experiments

The environment in which evolution occurs is extremely important to the final results. The simulated environment imposes similar constraints on the virtual humanoid robot as the natural environment would on a real humanoid robot replica. It

Table 2 Genetic Algorithm Parameters

Parameters	Value
Population size	100
Elitism	20 %
Crossover Rate	70 %
Genomic Mutation Rate	1%
Selection type	Roulette wheel selection
Chromosome length	1130 (45 neurons)
The number of generations	Up to 200

is likely that any real humanoid robot counterpart would encounter "obstacles" in the natural environment. Thus, in order to allow the humanoid robot to learn to cope with obstacles, they should be modelled within the virtual environment. A complex environment with obstacles that the humanoid robot will need to know in order to avoid (or make use of) may provide for robust evolved behaviors.

4.1 The Software

The software is coded in C++ without parallelization and the experiments were carried out on a desktop workstation. The creatures 3D environment and physics are simulated by Russel Smiths Open Dynamics Engine version 0.11.2[2] .

4.2 Robot Morphology

For our simulation, we used a robot humanoid constrained to move on a flat surface. The robot has 15 degrees of freedom: torques can be applied at both knee joints and at both hip joints. In addition, an additional actuator controls the posture of the upper body.

4.3 Physics Parameters

Different character parameters have been used. To simulate a structure in a physical way, there are parameters like bodies mass, bodies dimension, bodies centre of mass offset, frictions, joints limits and maximum forces applicable by joints that are very important to obtain realistic movements. Some parameters are related because, for example, the bodies mass, dimension and COM (Centre of mass) offset affect the maximum forces applicable by joints. The physics engine simulates a continuous world in an approximate and discrete way. Setting the max force applicable by joints, it is important to re-member that the physics engine is not always accurate. For example greater forces and torques, cause greater errors for accumulate

---

[2] Available at: http://ode.org/

during the physics simulation. This can happen because the physics engine manipulates quantities that span an increasing number of orders of magnitude with fixed-precision numbers. The physics engine errors can implicate the violation of the movements limits of a joint, and thus incurring, unrealistic motions.

Static and dynamic frictions are other important parameters to generate a realistic motion. The friction values are related to the other parameters because, naturally, forces applicable and masses affect the bodys reaction during the motion. Using little values implicates more difficulty to achieve a goal because the character cannot obtain the right push to walk. The tests about different friction values have been useful to understand better how several ground specifications can affect the human motion. However, you can choose what material the robot and the ground are made of, and look up the corresponding static/dynamic friction parameters.

5   Results

We here show the ability of our system to evolve a controller of a humanoid robot in several situations. Two test cases were used: locomotion control on flat terrain and obstacle avoidance. The two experiments have been recorded on videos[3] .

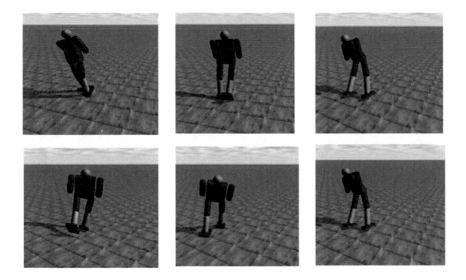

Fig. 5 A sequence of humanoid robot that is tryng to walk

---

[3] A video of our results can be viewed at: http://siva.univbiskra.net/Demos/humanoid_robot_project.wmv

5.1   Results of the Locomotion on Flat Terrain

Figure.5 shows some attempts to keep the robots balance, these images are from early generations so the movement is not displacement on this stage, because the robot is only moving on its place.

As fig. 6 shows, the robot can successfully walk on the flat terrain by periodically and alternatively stepping forward. However, it is observed that this evolved controller is unable to cope with slops. Note that at this stage, only the contact sensors were used in the feet.

Fig.7. Shows the oscillations of the hip-joint angle, knee-joint angle, ankle-joint angle and arm-joint angle of the left and right legs of the humanoid robot during locomotion relative a simulation period of 10 seconds.

The plots of each leg joint are them from the top individual in the best generation (that is generation 229).

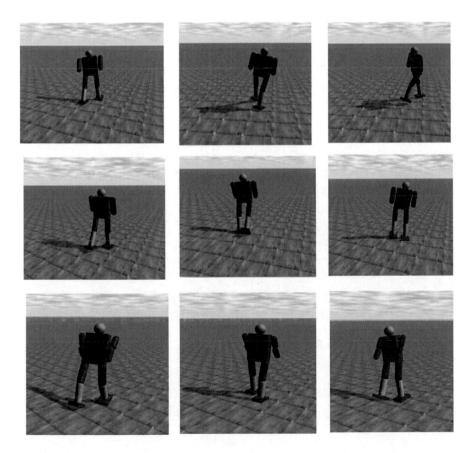

Fig. 6 The best evolved gait of humanoid robot that moves from right to left. This series of moves would be repeated over and over in a stable, natural gait.

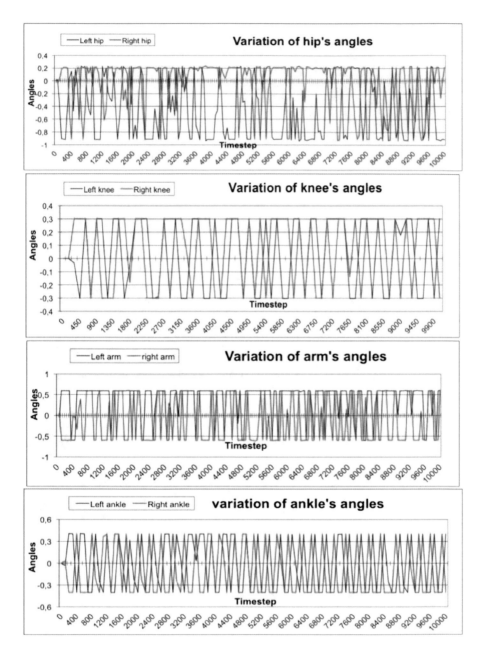

Fig. 7 Motor neuron activation levels of top individual in generation 229

By contrast, the other graphs represent symmetrical oscillations of the angles of the right and left feet. In general, we notice in the four graphs the oscillations from the right foot are almost opposite of those from the left foot

These graphs show that the robot succeeded in moving while advancing step by step in a non-periodic way (i.e. not simultaneously). This is due to the fact that in certain levels, the right foot is positioned according to its maximum angle but the left foots angle is on zero (null) value or the minimal value. This can be seen especially from the first graph of the variations of hips angles where the robot does not generate completely symmetrical movements. However, starting from the seventh second of the simulation, the movements of the hips become symmetrical.

The best evolved gait of humanoid robot that moves from right to left. This series of moves would be repeated over and over in a stable, natural gait.

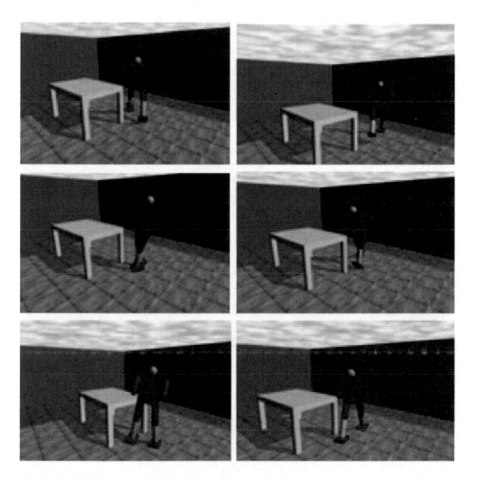

Fig. 8 Avoidance obstacle of the humanoid robot during locomotion

The robot can continue walking without falling over by modulating the torques applied at its feet.

5.2 Results of the Locomotion on Terrain with Obstacles

The robot can continue walking without falling over by modulating the torques applied at its feet. In this case study, two angles sensors are used in each foot to have more information about the obstacles.

When the humanoid robot detects obstacles during its course, it changes direction by swiveling its body to orientate itself forward the new direction. The best behavior obtained for obstacles avoidance is presented in fig. 8.

6 Discussion

The evolutionary process in this work was able to successfully produce a stable bipedal walking gait that allowed the humanoid robot to move forward throughout the complete duration of its evaluation period.

Fig.9 shows representative runs of the humanoid robot. Different distances are achieved with evaluated times (10 sec). The robot walking speed is accordingly higher.

Fig.9 shows also, average and maximum fitness values of the robot over 250 generations. Evolutionary runs took approximately two weeks of simulated time on a medium workstation. We noticed that for about 40 generations, no efficient gait emerges. Then, there is a nice steady increase in fitness up to about generation 140, followed by a jagged pattern with a steady upward trend. Any fitness above about

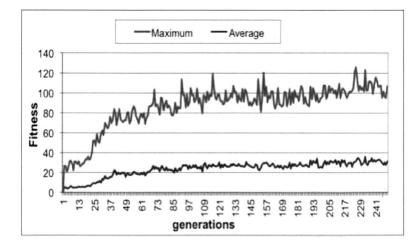

Fig. 9 Fitness graph of representative stable controller evolution. Top fitness and average fitness are shown.

the generation 200 will generally correspond to a reasonable walk in the forward direction, at generations 230 or above corresponding to quite good, stable walks. The maximum fitness generated was at generation 229, with a value of 93.76 corresponding to a fine forward walk with a slight limping gait.

Other patterns of locomotion included skating, with the robot keeping one foot constantly in contact with the ground and pushing along with the other.

7   Conclusion and Future Work

We have demonstrated the suitability of an evolutionary robotics approach to the problem of generating stable three-dimensional bi-pedal walking behaviors. The current implementation is able to evolve agents that are able to walk in a straight line on a planar surface without proprioceptive input. However, the use of proprioceptive sensors will become necessary to stabilize the biped on uneven terrain or in response to directional changes. Incorporation of such additional input is easy with the neural controller employed in this research.

The quality of the results is expected to be further improved by a refined fitness function, as well as a shift toward coupled neural oscillators instead of a single network. Furthermore, it is desirable to incorporate biomechanical knowledge about human walking in order to make maximum use of the passive dynamics of the bodies. These aspects are currently being implemented.

In theory, the results obtained here are directly transferable to em-bodied robots. In practice, however, there are likely to be complications due to a possible lack of accuracy of the physics engine. It re-mains to be seen whether this reality gap can be crossed with appropriate techniques such as noise envelopes [13].

Evolutionary algorithms typically use direct encodings, where each element of a phenotype is independently specified in its genotype. However, these direct encodings are limited in their ability to evolve complex, modular, and symmetric phenotypes because individual mutations cannot produce coordinated changes to multiple elements of a phenotype [11]. Such coordinated mutational effects can occur with indirect encodings, also called developmental or generative encodings, wherein a single element in a genotype can influence many parts of the phenotype [11]; [21]. Indirect encodings have been shown to produce highly regular solutions to problems [11]; [4]; [3] and [12], but their bias toward regularity makes it difficult for them to properly handle irregularities in problems [3].

Analyses suggest that HyperNEAT is successful because it em-ploys a generative encoding that can more easily reuse phenotypic modules. It is also one of the first neuroevolutionary algorithms that exploits a problem's geometric symmetries, which may aid its performance. [4]

For this reason it is important to do a comparative study of our system with the HyperNeat approach.

References

1. Azarbadegan, A., Broz, F., Nehaniv, C.L.: Evolving Simss Creatures for Bipedal Gait. In: IEEE Symposium on Artificial Life, Paris, France, April 11-15. Symposium series on computational intelligence, pp. 218–224 (2011)
2. Cheng, M.Y., Lin, C.S.: Genetic algorithm for control design of biped locomotion. Journal of Robotic Systems 14(5), 365–373 (1997)
3. Clune, J., Ofria, C., Pennock, R.T.: How a generative encoding fares as problem regularity decreases. In: Rudolph, G., Jansen, T., Lucas, S., Poloni, C., Beume, N. (eds.) PPSN 2008. LNCS, vol. 5199, pp. 358–367. Springer, Heidelberg (2008)
4. Clune, J., Beckmann, B.E., Ofria, C., Pennock, R.T.: Evolving coordinated quadruped gaits with the hyperneat generative encoding. In: IEEE Congress on Evolutionary Computing (CEC), Trondheim, Norway, pp. 2674–2771 (2009)
5. Doerschuk, P.I., Simon, W.E., Nguyen, V., Li, A.: A Modular Approach to Intelligent Control of a Simulated Jointed Leg. IEEE Robotics and Automation Magazine 5(2), 12–21 (1998)
6. Gallagher, J.C., Beer, R.D., Espenschied, K.S., Quinn, R.D.: Application of evolved locomotion controllers to a hexapod robot. Robotics and Autonomous Systems 19(1), 95–103 (1996)
7. Elman, L.J.: Finding Structure in Time. Cognitive Science 14, 179–211 (1990)
8. Gruau, F.: Automatic definition of modular neural networks. Adaptive Behavior 3(2), 151–183 (1995)
9. Haykin, S.: Neural Networks: A comprehensive foundation, 2nd edn. Prentice Hall, Upper Saddle River (1999)
10. Holland, J.: Adaptation in Natural and Artificial Systems. MIT Press, Cambridge (1992)
11. Hornby, G.S., Lipson, H., Pollack, J.B.: Generative representations for the automated design of modular physical robots. IEEE Transactions on Robotics and Automation 19, 703–719 (2003)
12. Hornby, G., Takamura, S., Tamamoto, T., Fujita, M.: Autonomous evolution of dynamic gaits with two quadruped robots. IEEE Transactions on Robotics 21(3), 402–410 (2005)
13. Jakobi, N., Husbands, P., Harvey, I.: Noise and the reality gap: The use of simulation in evolutionary robotics. In: Morán, F., Merelo, J.J., Moreno, A., Chacon, P. (eds.) ECAL 1995. LNCS, vol. 929, pp. 704–720. Springer, Heidelberg (1995)
14. Kun, A.L., Miller, W.T.: Control of variable speed gaits for a biped robot. IEEE Robotics and Automation Magazine 6(3), 19–29 (1999)
15. Liu, H., Iba, H.: A hierarchical approach for adaptive humanoid robot control. In: Proceedings of the 2004 IEEE Congress on Evolutionary Computation, Portland, Oregon, pp. 1546–1553, 20–23. IEEE Press, Los Alamitos (2004)
16. Miller III, W.T., Glanz, F.H., Kraft, L.G.: Application of a General Learning Algorithm to the Control of Robotic Manipulators. International Journal of Robotics Research 6(2), 84–98 (1987)
17. Miller, W.T.: Real-Time Neural Network Control of a Biped Walking Robot. IEEE Control Systems Magazine, 41–48 (1994)
18. Pettersson, J., Sandholt, H., Wahde, M.: A flexible evolutionary method for the generation and implementation of behaviors for humanoid robots. In: Proceedings of the IEEE-RAS International Conference on Humanoid Robotic, Japan, November 22-24, pp. 279–286 (2001)
19. Shan, J., Junshi, C., Jiapin, C.: Design of central pattern generator for humanoid robot walking based on multi-objective GA. In: Proc. International Conference on Intelligent Robots and Systems (IROS 2000), vol. 3, pp. 1930–1935. IEEE-RSJ, Takamatsu (2000)

20. Sims, K.: Evolving 3D morphology and behavior by competition. Artificial Life 1(4), 353–372 (1994)
21. Stanley, K.O., Miikkulainen, R.: A taxonomy for artificial embryogeny. Artificial Life 9(2), 93–130 (2003)
22. Taga, G., Yamaguchi, Y., Shimizu, H.: Self-organized control of bipedal locomotion by neural oscillators in unpredictable environment. Biological Cybernetics 65, 147–159 (1991)
23. Takanishi, A., Ishid, M., Yamazaki, Y., Kato, I.: The realization of dynamic walking by the biped walking robot WL-10RD. In: Proceedings of the International Conference on Advanced Robotics (ICAR 1985), pp. 459–466 (1985)
24. Téllez, R.A., Angulo, C., Pardo, D.E.: Evolving the walking behaviour of a 12 DOF quadruped using a distributed neural architecture. In: Ijspeert, A.J., Masuzawa, T., Kusumoto, S. (eds.) BioADIT 2006. LNCS, vol. 3853, pp. 5–19. Springer, Heidelberg (2006)
25. Valsalam, V.K., Miikkulainen, R.: Modular neuroevolution for multi-legged locomotion. In: GECCO 2008: Proceedings of the 10th annual conference on Genetic and Evolutionary Computation, pp. 265–272. ACM, New York (2008)
26. Wang, H., Lee, T.T., Gruver, W.A.: A neuromorphic controller for a three-link biped robot. IEEE Transactions on Systems, Man and Cybernetics 22(1), 164–169 (1992)
27. Wolff, K., Nordin, P.: Learning biped locomotion from first principles on a simulated humanoid robot using linear genetic programming. In: Cantú-Paz, E., Foster, J.A., Deb, K., Davis, L., Roy, R., O'Reilly, U.-M., Beyer, H.-G., Kendall, G., Wilson, S.W., Harman, M., Wegener, J., Dasgupta, D., Potter, M.A., Schultz, A., Dowsland, K.A., Jonoska, N., Miller, J., Standish, R.K. (eds.) GECCO 2003. LNCS, vol. 2723, pp. 495–506. Springer, Heidelberg (2003)
28. Ziegler, J., Barnholt, J., Busch, J., Banzhaf, W.: Automatic evolution of control programs for a small humanoid walking robot. In: Bidaud, P. (ed.) Proc. 5th International Conference on Climbing and Walking Robots (CLAWAR 2002), pp. 109–116. Professional Engineering Publishing (2002)

Techniques for Extracting and Modeling Geometric Features from Point Cloud Data Sets with Application to Urban Terrain Modeling

Gregory M. Nielson

Computer Science and Mathematics, Arizona State University,
Tempe, Arizona, USA
nielson@asu.edu

Abstract. Some new methods for extracting geometric features in point clouds are described. In addition new methods for including these geometric entities into implicit mathematical models are also discussed. Applications of these new techniques to the modeling of urban terrain data are illustrated.

Keywords: Scattered data, point clouds, urban terrain modeling, implicit mathematical models, scientific visualization.

1 Introduction and Motivation

We describe some new techniques for extracting geometric features from point cloud data sets. In addition, we discuss how to incorporate these geometric features into implicit mathematical models for fitting scattered, noisy point cloud data. One of the main applications we have in mind is virtual (see [4]) urban terrain environments consisting of both natural and manmade objects such as buildings, bridges and vehicles. See examples of urban terrains in Fig. 1 and note that conventional methods of fitting functions to this type of implied geometry does not apply due to the observation that urban terrain is not a function with respect to the surface of the Earth. So rather than use the conventional function methods (see [3], [7] and [10]) of modeling a point cloud $P = (x_i, y_i, z_i) \, i = 1, \cdots, N$ with a function F such that $z_i \approx F(x_i, y_i)$ we will use implicit methods where $F(x_i, y_i, z_i) \approx 0$ which means that the geometry of an urban environment is the level set of a trivariate function, that is, the set of points (x, y, z) such that $F(x, y, z) = 0$. This method of representing the geometry of an urban scene is particularly useful for many of the operations that are of interest. This includes not only rendering and simulation, but also the Boolean operation of adding or removing objects from the scene. A unique feature of the methods discussed here is the lack of the need for estimates of normal vectors which are required for most of the presently available methods (see [5], [6] and [8]). Normal vector estimation

D. Plemenos and G. Miaoulis (Eds.): Intelligent Comp. Graphics 2011, SCI 374, pp. 175–195.
springerlink.com   © Springer-Verlag Berlin Heidelberg 2012

Fig. 1 Examples of urban terrain containing natural objects and artifacts

adds to the computation cost of the computing a model and is problematic and often brings in errors and orientation issues early in the fitting process. Overall, it is considered to be an advantage to avoid the normal vector estimation process and this is accomplished with the present method. The present method is based upon least squares fitting of the scattered point cloud data and thus it is well suited for large and potentially noisy scattered point cloud data sets.

2 Implicit Mathematical Models for Scattered Point Clouds

We first describe our new method applied only to a collection of scattered data points $P_i, i = 1, \cdots, N$ and we will later explain how we incorporated other geometric objects, such as edges and facets, into the fitting process of determining the modeling function. We are interested to find a modeling function $F(P)$ so that the zero level contour represents an approximation to the data in the sense that $F(P_i) \approx 0, \; i = 1, \cdots, N$. If we let the basis functions for the modeling function be denoted by $B_j \quad j = 1, \cdots, M$ and we use the least squares error criteria then we have the basic fitting problem:

$$\min_{F_j} \sum_{i=1}^{N} \left[\sum_{j=1}^{M} F_j B_j (P_i) \right]^2 .$$

Without additional constraints, this problem is not well formulated since the optimal fit is the identically zero function. Many previous methods (see [5] and [6]) involve the estimation of normal vectors, N_i, and further requires that $\nabla F(P_i)$ approximate N_i. Here we do something rather different in that we add the additional constraints of normalizing the weights of the fitting function and consider the modified optimization problem

$$\min_{F_j} \left\{ \sum_{i=1}^{N} \left[\frac{\sum_{j=1}^{M} F_j B_j (P_i)}{\sqrt{\sum_{j=1}^{M} F_j^2}} \right]^2 \right\} \tag{1}$$

We now set out to find the characterizing equations for the minimizing set of coefficients, $F_j \quad j = 1, \cdots, M$. While it is possible to invoke some general results concerning Rayleigh quotients, it is both interesting and instructive to proceed using conventional calculus techniques. To this end, we introduce the following notation for the objective function of (1),

$$\phi(F_1, F_2, \cdots, F_M) = \frac{\sum_{i=1}^{N} \left[\sum_{j=1}^{M} F_j B_j (P_i) \right]^2}{\sum_{j=1}^{M} F_j^2} \,. \tag{2}$$

A minimizer of ϕ, necessarily must be a stationery point; that is, a root of the gradient, which requires that

$$\nabla \phi(F_1, F_2, \cdots, F_M) = 0 = \begin{pmatrix} \dfrac{\partial \phi(F_1, F_2, \cdots, F_M)}{\partial F_1} \\ \vdots \\ \dfrac{\partial \phi(F_1, F_2, \cdots, F_M)}{\partial F_M} \end{pmatrix}. \tag{3}$$

Using the standard calculus quotient rule for derivatives, we have

$$\frac{\partial \phi}{\partial F_k} = \frac{\|F\|^2 \left\{ \sum_{i=1}^{N} \left(2 \left[\sum_{j=1}^{M} F_j B_j (P_i) \right] B_K (P_i) \right) \right\} - \left\{ \sum_{i=1}^{N} \left[\sum_{j=1}^{M} F_j B_j (P_i) \right]^2 \right\} 2F_k}{\|F\|^4} = 0,$$

where $\|F\|^2 = \sum_{j=1}^{M} (F_j)^2$. It is easy to see that the above system of equations is equivalent to

$$A \begin{pmatrix} F_1 \\ F_2 \\ \vdots \\ F_M \end{pmatrix} - \frac{\sum_{i=1}^{N} \left[\sum_{j=1}^{M} F_j B_j (P_i) \right]^2}{\|F\|^2} \begin{pmatrix} F_1 \\ F_2 \\ \vdots \\ F_M \end{pmatrix} = 0, \tag{4}$$

where $A = BB^*$ denotes the gram matrix and

$$B = \begin{pmatrix} B_1(P_1) & B_1(P_2) & \cdots & B_1(P_N) \\ B_2(P_1) & B_2(P_2) & \cdots & B_2(P_N) \\ \vdots & \vdots & \ddots & \vdots \\ B_M(P_1) & B_M(P_2) & \cdots & B_M(P_N) \end{pmatrix}. \tag{5}$$

At this point, we assume that the data, $P_i, i = 1, \cdots, N$ and the basis functions $B_j, j = 1, \cdots, M$ are such that A is positive definite. From Equation (4), we can see that minimizing solution, F, must be an eigenvector of A. If we let λ be the associated eigenvalue then we may also conclude that

$$\frac{\sum_{i=1}^{N} \left[\sum_{j=1}^{M} F_j B_j (P_i) \right]^2}{\sum_{j=1}^{M} F_j^2} = \lambda \tag{6}$$

and so we have established the fact that the vector F that minimizes (2) is the eigenvector of the gram matrix, BB^* associated with its smallest eigenvalue. We now present a very simple example which illustrates this new fitting process. In addition, this example can be subsequently used by others to check and verify implementations and basic conceptual issues of our new method. This example uses only 2D data as opposed to the real world situation where the point cloud is 3D but the example will explain the basic ideas. For this example we have thirteen (13) data points contained in the domain consisting of the unit square, $(.44,.13)$, $(.24,.15)$, $(.22,.35)$, $(.31,.49)$, $(.35,.60)$, $(.43,.70)$, $(.51,.77)$, $(.57,.70)$, $(.70,.55)$, $(.85,.30)$, $(.88,.23)$, $(.80,.10)$ and $(.60,.10)$.

The implicit modeling function, $F(x, y)$ is defined to be piecewise linear function defined over the four triangles with vertices at the four corners plus the center point $(.5,.5)$. The data and the contour of the optimal fit is shown in the top image of Fig. 2 and the image of Fig. 3 shows the height field rendering of the optimal implicit model which is defined by the values:

$$F(0,1) = -.6750, \quad F(1,1) = -.5571,$$
$$F(.5,.5) = +.4281, \quad F(0,0) = -.2142,$$
$$F(1,0) = -.0694$$

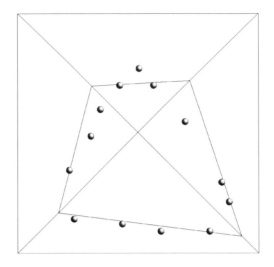

Fig. 2 The data input for a simple example to illustrate the main ideas of the general method of normalized implicit eigenvector least squares operators for noisy scattered data. The image shows the 13 data points and the 4 triangles which comprise the domain

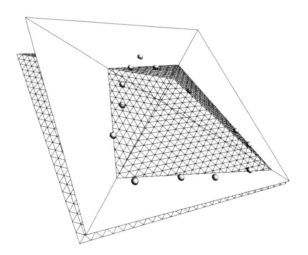

Fig. 3 Thid image shows the piecewise linear implicit fit to the data consisting of 13 scattered points. Note that no normal vector estimates are involved in the approximation

The next novel aspect of our new modeling technique is how we incorporate edges and facets into the modeling function. Again, we can explain the basic ideas within the context of 2D input data. See the data of Fig. 4 where we have scattered points along with line segments and polygons as input data.

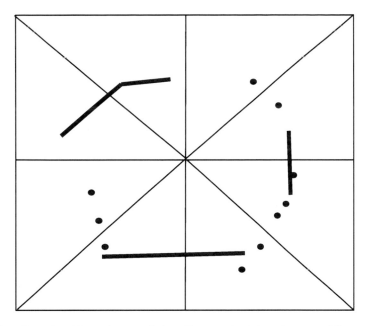

Fig. 4 Two dimensional input data consisting of scattered points, polygons and line segments

We will explain our techniques for the case of piecewise linear implicit models of 2D and the generalization to higher order models (trivariate, quadratic) in 3D will be clear.

For a piecewise implicit model in 2D we use barycentric coordinate to represent the model locally. The first step consists of clipping all input geometry to the triangle (tetrahedral) cells of the domain of the fitting function F. Previously, we had the fitting problem

$$\min_{F_j} \sum_{i=1}^{N} \left[\sum_{j=1}^{M} F_j B_j (P_i) \right]^2 \tag{7}$$

where

$$C(F_1, \cdots, F_N) = \sum_{i=1}^{N} \left[\sum_{j=1}^{M} F_j B_j (P_i) \right]^2$$

is called the cost function and the unknowns, computed during the optimization process consist of $F_1, F_2, \cdots, F_N$. In this 2D situation, each scattered input point P_n adds the following to the cost function

Cost function = Cost function + $\left[F_i b_i (P_n) + F_j b_j (P_n) + F_k b_k (P_n) \right]^2$

where $\left[b_i(P_n), b_j(P_n), b_k(P_n)\right]$ is the barycentric coordinate of the point P_n, in the triangle that contains this point. Each trimmed line segment (triangle facet in the 3D context) adds the following to the cost function

Cost function = Cost function +

$$\int_0^1 \left[\begin{array}{l} F_i b_i (tp_b + (1-t)p_e) + F_i b_i (tp_b + \\ (1-t)p_e) + F_i b_i (tp_b + (1-t)p_e) \end{array}\right]^2 dx$$

We now note that the integrand of the cost function contribution for a triangle facet is a second order polynomial and so we can replace the entire integrand with a sum of terms like the point contribution using any quadrature rule that is precise for quadratic polynomials. For example, in the 2D version with trimmed line segments, we could use Simpson's rule which would involve the points

$$w_1\left\{\left[F_i b_i(p_b) + F_j b_j(p_b) + F_k b_k(p_b)\right]^2\right\} +$$

$$w_2\left\{\left[\begin{array}{l} \left[F_i b_i\left(\dfrac{p_b + p_e}{2}\right) + F_j b_j\left(\dfrac{p_b + p_e}{2}\right)\right] \\ + F_k b_k\left(\dfrac{p_b + p_e}{2}\right) \end{array}\right]^2\right\}$$

$$+ w_3\left\{\left[F_i b_i(p_e) + F_j b_j(p_e) + F_k b_k(p_e)\right]^2\right\}$$

With quadrature rules replacing the integrals, the entire cost function consists of the sum of similar terms involving only point evaluation squared. Next we look at the weight functions for each of the F_i. Each of these consists of the sum of point evaluation at scattered points or at points on the facets. The important point to note is that by the use of numerical quadrature methods, the introduction of triangular facets into the cost function does not change the basic form of the overall minimization problem that must be solved from that given in (1) above. This means that eigenvector/eigenvalues as before are the optimal solutions for the implicit model!

In the previous discussion we have shown that involving line segemts into the overall error model by means of integral measure error does not change the overall structure of the model to be minimized as long as the integrals are replaced with quadrature approximations which in this case are chosen so as to be exact (quadrature rules). A corner is involved in the same manner, but with two integrals which are replaced with quadrature approximations (which are exact with no error).

Involving clipped facets consisting of triangles into the 3D model is done in exactly the same manner.

3 The Nonlinear Error Weighted Eigen (NEWE) Method of Feature Detection

Corner and edge detection from point clouds is a very desirable capability for many applications. Currently, it is also a very difficult problem with very few effective or useful techniques. Using discrete L_p norms, nonlinear optimization techniques and techniques related to the adaptive implicit fitting discussed above, we have recently developed a general approach and class of methods for analyzing the data yielding the positions of these corners. Unlike most of the work in computer vision which relates to this problem, our methods directly extend to any number of dimensions and are based upon geometric modeling concepts. We have written some demonstration programs for 2D and 3D which allow a user to test our techniques on some examples. The images of Fig. 5, Fig. 6, and Fig. 7 show typical screen shots of one of these programs in action. The demonstration programs can be downloaded (www.public.asu.edu/~nielson/ARO/) which add additionally insight into how the methods are capable of being so effective.

We first explain the basic ideas of these new methods within the context of the simplest possible case of two lines in 2D. The method is based upon the iterative use of the eigenvector/value method of optimized implicit fitting described earlier in this paper.

Initialize : (A, B, C) as the solution to

$$\underset{A,B,C}{Min} \frac{\sum_i (Ax_i + By_i + Cz_i)^2}{A^2 + B^2 + C^2} \tag{8}$$

For $P = P_1, P_2, \cdots, P_n$ solve (for both (9) & (10))

$$\underset{a,b,c}{Min} \frac{\sum_i (ax_i + by_i + cz_i)^P (Ax_i + By_i + Cz_i)^2}{a^2 + b^2 + c^2} \tag{9}$$

followed by

$$\underset{A,B,C}{Min} \frac{\sum_i (ax_i + by_i + cz_i)^2 (Ax_i + By_i + Cz_i)^P}{A^2 + B^2 + C^2} \tag{10}$$

Often, $P_i = 2^{n+1-i}$ with $n = 5$, but other choices are possible and interesting.

It is clear how to extend this iterative technique of using the error for one fit as the weights in the next fit to any number of lines or planes.

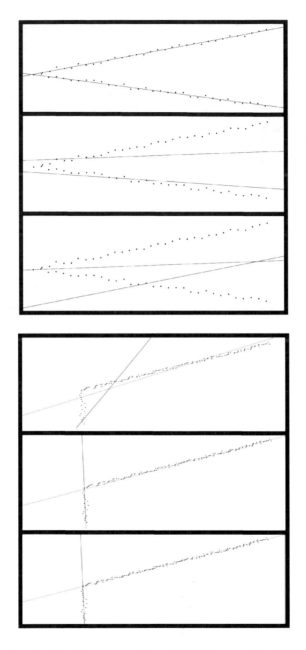

Fig. 5 Two examples showing (from top to bottom) the first, second and final iterations of the new iterative method of determining corners implied by scattered point cloud data. The method is based upon the discrete Lp norms and nonlinear optimization. One example is shown in the top three images and another is shown in the bottom three images

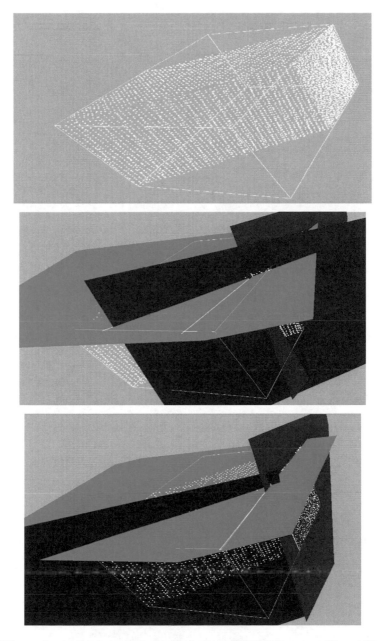

Fig. 6 The top image shows a point cloud inferring an "outside" corner. The middle image shows the iterative algorithm in action and the bottom image shows the final converged result modeled with planes clipped and triangulated and added to the input data for the implicit modeling algorithm

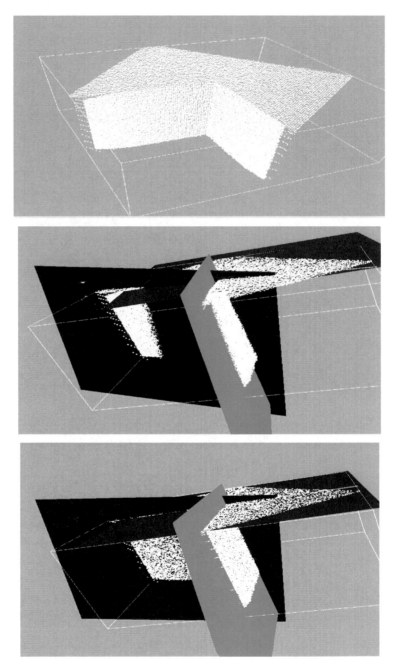

Fig. 7 The top image shows a point cloud inferring a "knotch" corner. The middle image during iteration and the bottom image shows the final converged result

Fig. 8 The top image shows the input data points. The middle is the piecewise quadratic implicit model which captures the corner feature automatically and accurately. The bottom image shows the inadequate properties of the piecewise linear implicit model

4 Discussion, Rendering Techniques and Examples

We now want to point out one of the most important key features of this type of
implicit modeling, which is the ability to efficiently and effectively model sharp
features implied by the data. The ability to do this is a direct consequence of the
choice for the implicit field function. A simple example is shown in Fig. 8. In
this case, the modeling function is a piecewise, implicit quadratic polynomial
which can have a hyperbola (and it degenerate case of two intersecting lines) as its
level set. It is really rather remarkable how well this process works. This entire
point cloud is fit (just about perfectly!) with only a small number of implicit
curves. This example also points out one potentially undesirable attribute of this
type of model and this is the possible appearance of extraneous contours. But this
is not a real problem as these extraneous contours are easily removed by using the
context of data in cells and continuity of contours. In order for a piecewise linear
model (as opposed to quadratic) to fit a corner, the corner would have to be on the
boundary between domain cells. This means, if we use quadratic basis functions,
we do not have to know the location of the corner in advance and involve this in
the definition of the fitting function.

We now show three 3D examples which illustrate the power of this class of implicit methods to model point cloud data. They are rather different from each other. The first is more traditional and consist of a mechanical part (Fig. 9 and Fig.
10). The second two illustrate the utility of this class of methods for modeling urban terrain data consisting of natural terrain and artifacts (Fig. 12 and Fig. 13).
One of the reasons for including the first example is to underscore the ease of
computing Boolean operations (see [1]) on models that are implicitly defined as is
the case for the fitting models of this paper. Let F_A denote the field function for
a point set A whose boundary is a surface of interest. That is
$A = \left\{ (x, y, z) : F_A(x, y, z) \geq 0 \right\}$. If B is another three dimensional point set
with the field function F_B, then it is easy to see that the union is defined by

$$A \cup B = \left\{ (x, y, z) : Max(F_A(x, y, z), F_B(x, y, z)) \right\}$$

and so we have that $F_{A \cup B} = Max(F_A, F_B)$. Similarly, we have for the intersec-
tion $F_{A \cap B} = Min(F_A, F_B)$.

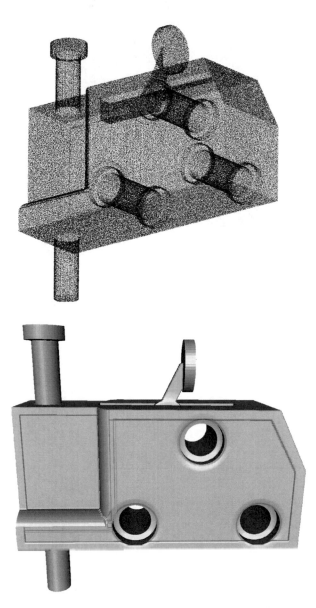

Fig. 9 The top image shows the input data points scanned from a patio door lock. The bottom is level set of the implicit model

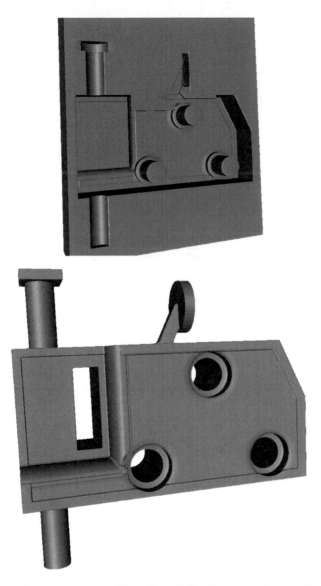

Fig. 10 The top image shows a mold made with Boolean operations applied to the field function and the bottom images shows some additional modifications made with Boolen operations: a notch has been cut out and a square top has been put on the latch pin.

The rendering of the isosurfaces shown in Fig. 9, Fig. 10, Fig. 12 and Fig. 13 are based upon triangular mesh surfaces which are extracted from the implicit model using the Dual Marching Tetrahedra (DMT) method described in [9]. We use this "Dual" method so that the corner and edge features are maintained through the isosurface extraction process and do not necessarily lie on cell domain boundaries. For completeness, we briefly describe the DMT method here.

Let $P_i = (x_i, y_i, z_i), i = 1, \cdots, N$ denote the grid points which are segmented as marked or unmarked. We assume these points are not collectively coplanar. We assume that the grid points have been arranged into a collection of tetrahera to form a tetrahedronal. A tetrahedonal consists of a list of 4-tuples which we denote by I_t. Each 4-tuple, $ijkl \in I_t$ denotes a single tetrahedron with the four vertices P_i, P_j, P_k, P_l which is denoted as T_{ijkl}. A valid tetrahedronal requires: i) No tetrahedron $T_{ijkl}, ijkl \in I_t$ is degenerate, i. e. the points P_i, P_j, P_k, P_l are not coplanar, ii) The interiors of any two tetrahedra do not intersect and iii) The boundary of two tetrahedra can intersect only at a common triangular face.

A tetrahedron is said to be active if among its 4 grid points there are both marked grid points and unmarked grid points. There are three distinct configurations for these active tetrahedra as shown in Figure 11. Similarly, a triangular face of an active tetrahedron is active provided it contains both marked and unmarked grid points. Interior to each active tetrahedron, $T_{n_i n_j n_k n_i}$, there is a vertex $V_{n_i n_j n_k n_i}$. For each interior active triangular face, there is an edge of S joining the two vertices of the two tetrahedra sharing this triangular face. If $F_{n_i n_j n_k}$ denotes the interior active triangular face and $T_{n_i n_j n_k n_a}$ and T_{n_i, n_j, n_k, n_b} denote the two active tetrahedra sharing this face then an edge joins $V_{n_i n_j n_k n_a}$ and $V_{n_i n_j n_k n_b}$.

We propose a scheme for computing the positions of the vertices within active tetrahedra that is based upon minimizing discrete norm curvature estimates at each vertex which we now describe. For the tetrahedron defined by the points P_i, P_j, P_k, P_l we consider the tetrahedra lattice points $V_{a,b,c,d} = (aP_i + bP_j + cP_k + dP_l)/N$ where the integers a, b, c, d satisfy the two conditions $0 < a, b, c, d < N$ and $a + b + c + d = N$. For each point $V_{a,b,c,d}$ in $T_{i,j,k,l}$ we use the discrete curvature method to compute an estimate of the norm curvature $k_1^2(S) + k_2^2(S) = 4[M(S)]^2 - 2K(S)$ where $k_1(S)$ and $k_2(S)$ are the principal curvatures, $M(S)$ is the mean curvature and $K(S)$ is Gaussian curvature. These estimates are based upon a triangulation of $V_{a,b,c,d}$ and the vertices of its 1-ring that maintain the edges of separationg surface S and do not introduce any additional edges containing $V_{a,b,c,d}$. The estimates are computed as $K(S) = 3(2\pi - \sum \alpha_i)/A$, $M(S) = .75 \sum \|e_i\| \beta_i / A$, where e_i is an edge joining $V_{a,b,c,d}$ and a vertex of its 1-ring, β_i is its dihedral angle,

α_i is a subtended angle and A is the sum of the areas of all the adjacent triangles.

We take as our first approximation $V_{i,j,k,l}^{(1)}$ the point $V_{a,b,c,d}$ associated with the smallest estimate of norm curvature. These values are computed for all tetrahedra containing vertices of the separating surface S. We do another pass over all of the tetrahedra containing vertices of S leading to the approximations $V_{i,j,k,l}^{(2)}$.

This is continued until the user specified criteria for convergence is satisfied. In practice usually 7 or 8 digits of accuracy are obtained in less than 6 iterations (a complete loop through all active tetrahedra). A resolution of N = 5, ... , 9 is a typical choice.

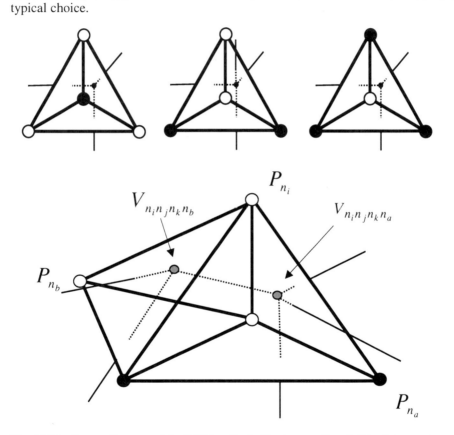

Fig. 11 The three active case of the DMT method. The top row from left to righ: one, two and three points classified as being contained in the object of interest. The lower image illustrates the notation of the tetrahedral method

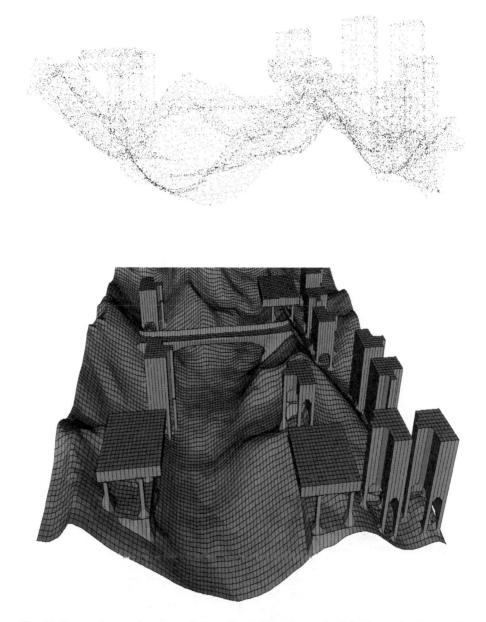

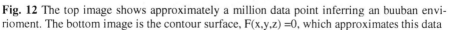

Fig. 12 The top image shows approximately a million data point inferring an buuban envirioment. The bottom image is the contour surface, $F(x,y,z) = 0$, which approximates this data

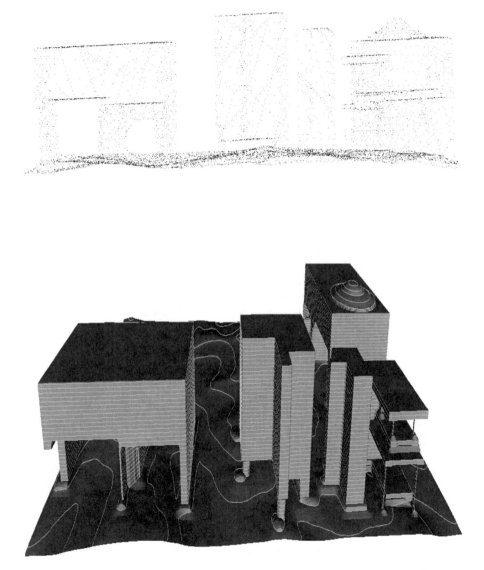

Fig. 13 The top image shows a scattered point cloud consisting of approximately one million data points representing an urban environment consisting of artifacts and natural terrain. The bottom image shows the levelset of the implicit mathematical equation which fits this data. Note that both sharp and smooth features can be modeled with a single implicit function

References

1. Bloomenthal, J., Bajaj, C., Blinn, J., Cani-Gascuel, M., Rockwood, A., Wyvill, B., Wyvill, G.: Introduction to Implicit Surfaces. Morgan Kaufmann, San Francisco (1999)
2. Cureless, G., Levoy, M.: A volumetric method for building complex models from range images. In: Proceedings of SIGGRAPH 1996. Annual Conference Series, pp. 303–312. ACM Press/ACM SIGGRAPH, New York (1996)
3. Franke, R., Nielson, G.M.: Scattered data interpolation and applications: A tutorial and survey. In: Hagen, H., Roller, D. (eds.) Geometric Modelling: Methods and Their Applications, pp. 131–160 (1990)
4. Klimenko, S., Nikitin, I., Burkin, V.: Visualization of Complex Physical Phenomena and Mathematical Objects in Virtual Environment. In: Scientific Visualization, Dagstuhl 1997, pp. 151–161. IEEE Computer Society Press, Germany (1999)
5. Mitra, N., Nguyen, A., Guibas, L.: Estimating surface normals in noisy point cloud data. International Journal of Computation Geometry & Applications, 123–145 (2004)
6. Mueller, H.: Surface Reconstruction – An introduction. In: Hagen, H., Nielson, G., Post, F. (eds.) Scientific Visualization, pp. 239–242. IEEE Computer Society Press, Germany (1990)
7. Nielson, G.: Scattered data modeling. Computer Graphics and Applications 13, 60–70 (1993)
8. Nielson, G.: Normalized implicit eigenvector least squares operations for noisy scattered data: radial basis functions. Computer 86(2-3), 199–212 (2009)
9. Nielson, G.: Dual Marching Tetrahedra: Contouring in the Tegrhedronal Environment. In: Bebis, G., Boyle, R., Parvin, B., Koracin, D., Remagnino, P., Porikli, F., Peters, J., Klosowski, J., Arns, L., Chun, Y.K., Rhyne, T.-M., Monroe, L. (eds.) ISVC 2008, Part I. LNCS, vol. 5358, pp. 183–194. Springer, Heidelberg (2008)
10. Schumaker, L.: Fitting surfaces to scattered data. In: Lorentz, G., Chui, C.K., Schumaker, L.L. (eds.) Approximation Theroy II, pp. 203–268 (1976)

Author Index